Lecture Notes in Social Networks

More information about this series at http://www.springer.com/series/8768

Mehmet Kaya • Özcan Erdoğan • Jon Rokne
Editors

From Social Data Mining and Analysis to Prediction and Community Detection

 Springer

Editors
Mehmet Kaya
Department of Computer Engineering
Firat University
Elazig, Turkey

Özcan Erdoğan
Ministry of Interior
Ankara, Turkey

Jon Rokne
Department of Computer Science
University of Calgary
Calgary, AB, Canada

ISSN 2190-5428   ISSN 2190-5436   (electronic)
Lecture Notes in Social Networks
ISBN 978-3-319-84631-6   ISBN 978-3-319-51367-6   (eBook)
DOI 10.1007/978-3-319-51367-6

Printed on acid-free paper

This Springer imprint is published by Springer Nature
The registered company is Springer International Publishing AG
The registered company address is: Gewerbestrasse 11, 6330 Cham, Switzerland

Preface

Introduction

This volume is a compilation of the best papers presented at the IEEE/ACM International Conference on Advances in Social Networks Analysis and Mining (ASONAM'2015), held in Paris, France, August 2015. The authors of these papers were asked to provide extended versions of the papers that were then subjected to an additional refereeing process. Within the broader context of online social networks, the volume focuses on important and upcoming topics such as attempting to understand the context of messages propagating in social networks, classifying sentiments and defining and understanding local communities.

From Social Data Mining and Analysis to Prediction and Community Detection

The importance of social networks in today's society cannot be underestimated. Millions of individuals use social networks to communicate with friends every day, and for many people (especially young ones) the social network presence is of vital importance for their identity.

Social networks are a rich source of information, and several of the papers in this volume focus on how to extract this information. The information gleaned from social networks can inform businesses, governments, social agencies, cultural groups and so on.

Social networks lead to questions such as privacy, ethics and data ownership. These questions and other issues are a fertile ground for further research.

The first paper in this volume is "An Offline-Online Visual Framework for Clustering Memes in Social Media" by Anh Dang, Abidalrahman Moh'd, Anatoliy Gruzd, Evangelos Milios and Rosane Minghim. The paper discusses the dissemination and clustering of memes in social networks where memes are "an element of a culture or system of behaviour that may be considered to be passed from one individual to another by non-genetic means, especially imitation or a humorous image, video, piece of text, etc., that is copied and spread rapidly by Internet users" (Bing definition). It turns out that social networks are ideal for the dissemination of memes and it is therefore interesting to study how the memes propagate through such networks and how the memes cluster in the networks. The authors compute similarity scores between texts containing memes and assess cluster memberships depending on the scores. The social network Reddit is studied in detail, and the clustering of memes in this network is used to detect emerging events. Experimental results show that their method implemented using the Google Trigram Method provides reasonable results.

Given that more than 200 billion e-mail messages are sent each day, it is to be expected that some messages will be sent to the wrong address. Most of us are likely guilty of doing it. Sometimes this can have serious consequences especially if sensitive information is transmitted. The paper by Zvi Sofershtein and Sara Cohen entitled "A System for Email Recipient Prediction" effectively predicts e-mail recipients when an e-mail history is available. This system takes into account a variety of clues from e-mail histories to predict recipients when there is sufficient data. The paper proposes a system based on a set of features and assesses the contribution to the precision of the system for each of the features. The system is tested on data sets and on various domains such as the Enron data set, a political data, a Gmail-English data set, a Gmail-Hebrew data set and a combined data set. Properly applied their system can reduce the number of unfortunate misdirected messages and increase the speed by which correctly addressed e-mails are composed.

Authenticating and verifying content of messages is common to many activities in today's world. Verified messages can be beneficial, whereas messages containing false information can be harmful (witness the scandal surrounding the falsified messages relating to the health benefits of sugar versus fat). Both true and false messages can be disseminated using online social networks. In paper "A Credibility Assessment Model for Online Social Network Content", Majed Alrubaian, Muhammad Al-Qurishi, Mabrook Al-Rakhami and Atif Alamri present an algorithmic approach for assessing the credibility of Twitter messages. They report encouraging results being able to achieve accuracies in the 80% range on two specific data sets.

Whereas the previous paper assessed the credibility of Twitter messages in Arabic, the paper by Mohammed Bekkali and Abdelmonaime Lachkar entitled "Web Search Engine Based Representation for Arabic Tweets Categorization" attempts to enhance Tweets (or in general short length messages) composed in Arabic by providing contextual information. They note that there is scant research in machine understanding of Arabic, partly due to its structure being different from the more studied European languages, and the understanding of short messages in Twitter in Arabic is therefore a twofold problem of both encoding (i.e. language) and the choice of language itself.

Preserving and increasing capital by investing in stocks is fraught with uncertainty. Reducing this uncertainty by applying sentiment analysis is discussed in the paper "Sentiment Trends and Classifying Stocks Using P-Trees" by Arijit Chatterjee and William Perrizo. The authors mine the large volume of Tweets generated each day for sentiment trends for specific stock symbols. They use Twitter API and P-trees to gain insight into what ticker symbols are "hot" and hence what ticker symbols and businesses should be a reasonable investment. Stock prices and sentiments are graphed for the two stock symbols AAPL (Apple) and FB (Facebook), and they conclude that the trends of stock prices follow the movements of the sentiments in a general sense.

Social networks in their simplest form consist of nodes and links between nodes. Essentially the links are the minimal information needed to form such networks. If the nodes contain richer information, then further insights can be gained from the networks. In their paper "Mining Community Structure with Node Embeddings", Thuy Vu and D. Stott Parker show that embeddings can reflect community structure. Node embeddings were implemented for the DLBP network, and it was shown that they were useful in finding communities.

Finding local communities in a social network has a number of applications. In the paper "A LexDFS-Based Approach on Finding Compact Communities", Jean Creusefond, Thomas Largillier and Sylvain Peyronnet apply an efficient graph transversal algorithm to identify clusters and hence communities in a social network. The paper proposes a new measure for connectedness. In their definition a subset of the social network is said to be connected if the distances between the nodes in the subset are small. They apply the LexDFS algorithm to compute connected communities and compare the results with six previously developed clustering algorithms.

Societies need laws and regulations to function effectively and to limit the damage unhinged individuals and entities can do if left unchecked. Laws and regulations therefore have a long history dating back to, at least, the Babylonians. Past laws and regulations are often modified and built upon to take into account changes and new advances in society, such as the Internet. Especially due to the Internet, the laws pertaining to privacy and data ownership have still to be defined in an acceptable form. In the past creating and modifying laws and regulations was the domain of well-paid legal experts. As society has become more and more complex and with the global reach of modern economies, the understanding and applying of laws has become more and more difficult and informatics tools are now applied to legal work. The financial sector is a fundamental and complex part of modern society. While it is the focus of much criticism, it is also an essential component of modern society. The aim of laws and regulations for the financial sector is to ensure a smooth functioning of economies while limiting excesses such as gambling with toxic derivatives (and collection obscene rewards for doing so). The resulting corpus of laws and regulations is very large and complex (especially in the United States of America), and it is difficult to get an overview of the corpus even for the experts. The paper "Computational Data Sciences and the Regulation of Banking and Financial Services" by Sharyn O'Halloran, Marion Dumas, Sameer Maskey, Geraldine McAllister and David Park attempts to alleviate some of the heavy lifting in this area by applying sophisticated computational algorithms to the corpus.

The final paper of the volume "Frequent and Non-frequent Sequential Itemsets Detection" by Konstantinos F. Xylogiannopoulos, Panagiotis Karampelas and Reda Alhajj considers the problem of detecting repeated patterns in a string. These patterns are classified as frequent sequential itemsets if they pertain to sets of events that are ordered. The proposed algorithm is based on transforming a set of transactions into a suitable data structure that then is processed using a novel process. This process was used for experiments with two data sets. The first data set was the "Retail" data set consisting of 88,162 transactions, and the second data set was the "Kosavak" data set with 990,002 transactions. In both cases all the frequent data sets were detected regardless of support value.

To conclude this preface, we would like to thank the authors who submitted papers and the reviewers who provided detailed constructive reports which improved the quality of the papers. Various people from Springer deserve large credit for their help and support in all the issues related to publishing this book.

Elazig, Turkey   Mehmet Kaya
Ankara, Turkey   Özcan Erdoğan
Calgary, AB, Canada   Jon Rokne
October 2016

Contents

An Offline–Online Visual Framework for Clustering Memes in Social Media

Anh Dang, Abidalrahman Moh'd, Anatoliy Gruzd, Evangelos Milios, and Rosane Minghim

1 Introduction

Online Social Networks (OSNs) are networks of online interactions and relationships that are formed and maintained through various social networking sites such as Facebook, LinkedIn, Reddit, and Twitter. Nowadays, hundreds of millions of people and organizations turn to OSNs to interact with one another, share information, and connect with friends and strangers. OSNs have been especially useful for disseminating information in the context of political campaigning, news reporting, marketing, and entertainment [32].

OSNs have been recently used as an effective source for end users to know about breaking-news or emerging memes. A meme is a unit of information that can be passed from person to person in OSNs [29]. Despite their usefulness and popularity, OSNs also have a "negative" side. As well as spreading credible information, OSNs can also spread rumours, which are truth-unverifiable statements. For example, so many rumour-driven memes about swine flu outbreak (e.g., "swine flu pandemic meme" in Fig. 1) were communicated via OSNs in 2009 that the US government

A. Dang (✉) • A. Moh'd • E. Milios
Faculty of Computer Science, Dalhousie University, 6050 University Avenue, PO BOX 15000, Halifax, NS, Canada B3H 4R2
e-mail: anh@cs.dal.ca; amohd@cs.dal.ca; eem@cs.dal.ca

A. Gruzd
Ted Rogers School of Management, Ryerson University, 55 Dundas Street West, Toronto, ON, Canada M5G 2C3
e-mail: gruzd@ryerson.ca

R. Minghim
University of São Paulo-USP, ICMC, São Carlos, Brazil
e-mail: rminghim@icmc.usp.br

© Springer International Publishing AG 2017
M. Kaya et al. (eds.), *From Social Data Mining and Analysis to Prediction and Community Detection*, Lecture Notes in Social Networks,
DOI 10.1007/978-3-319-51367-6_1

Fig. 1 A word cloud example of popular memes in OSNs

had to tackle it officially on their website [22, 37]. Problems like these (i.e., rumour-driven memes going viral) are unfortunately not isolated and prompt the question of how to identify and limit the spread of rumours in OSNs. In order to detect rumours, we have to identify memes that are rumour-related in OSNs. Clustering is a simple and efficient unsupervised process to identify memes in OSNs by grouping similar information into the same category. However, traditional clustering algorithms do not work effectively in OSNs due to the heterogeneous nature of social network data [31]. Labelling massive amounts of social network data is an intensive task for classification. To overcome these limitations, this paper proposes a semi-supervised approach with relevance user feedback for detecting the spread of memes in OSNs.

In text clustering, a similarity measure is a function that assigns a score to a pair of texts in a corpus that shows how similar the two texts are. Computing similarity scores between texts is one of the most computationally intensive and important steps for producing a good clustering result [6]. For a meme clustering task, this process is usually hindered by the lack of significant amounts of textual content, which is an intrinsic characteristic of OSNs [17, 27, 31]. For example, in Reddit.com, most submission titles are very short and concise. Although the title of a submission may provide meaningful information about the topic, the titles may not provide enough information to determine if two submissions are discussing the same topic. In Fig. 2, two Reddit submissions are both talking about "Obama," but one is discussing the meme "Obamacare," while the other is discussing the rumour-related meme "Obama is a Muslim." The sparsity of Reddit submission title texts significantly contributes to the poor performance of traditional text clustering techniques for grouping submissions into the same category. We, therefore, propose strategies to leverage the use of references to external content.

A submission may include one or more comments from users, which discuss the submission topic. It can also contain a URL that points to an external article that further discusses the topic of the submission. Similarly, a submission may include an image that also provides more valuable information about the submission topic. By introducing the use of comments, URL content, and image content of a submission, we exploit more valuable data for text clustering tasks, which helps detect memes in OSNs more efficiently.

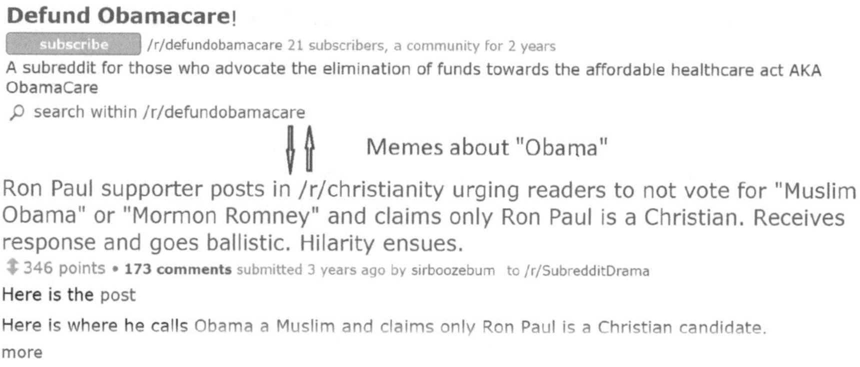

Defund Obamacare!

subscribe   /r/defundobamacare 21 subscribers, a community for 2 years

A subreddit for those who advocate the elimination of funds towards the affordable healthcare act AKA ObamaCare

🔎 search within /r/defundobamacare

⇅⇅ Memes about "Obama"

Ron Paul supporter posts in /r/christianity urging readers to not vote for "Muslim Obama" or "Mormon Romney" and claims only Ron Paul is a Christian. Receives response and goes ballistic. Hilarity ensues.

⬆ 346 points • **173 comments** submitted 3 years ago by sirboozebum to /r/SubredditDrama

Here is the post

Here is where he calls Obama a Muslim and claims only Ron Paul is a Christian candidate.

more

Fig. 2 Reddit submissions about the same meme "Obama." The *top submission* discusses the meme "Obamacare." while the *bottom submission* discusses the meme "Obama is a Muslim"

Vector space models are commonly used to represent texts for a clustering task. In these models, each text is represented as a vector where each element corresponds to an attribute extracted from the text. One of the benefits of these models is their simplicity in calculating the similarity between two vectors based on linear algebra. The two most famous models are TF-IDF (Term Frequency—Inverse Term Frequency) and Bag-of-Words. However, those models rely solely on lexical representation of the texts, which does not capture semantic relatedness between words in a text. For example, the use of polysemy and synonymy are very popular in several types of texts and play an important role in determining whether two words, concepts or texts are semantically similar. This motivates many researchers to explore the advantage of semantic similarity in the task of text clustering by utilizing word relatedness through external thesauruses like WordNet and Wikipedia [28, 38]. However, they remain far from covering every word and concept used in OSNs. This paper explores Google n-grams algorithm of Islam et al. [30] which uses Google n-grams dataset to compute the relatedness between words for computing similarity scores, and proposes two novel strategies to combine those scores for the task of clustering memes.

Although clustering streaming and time series data are established fields in the area of text clustering [14, 23–25, 41], clustering memes in OSNs has just started to gain attention recently [2, 3, 31, 42]. OSN data has both characteristics of streaming and time series data, as well as another important characteristic. The volume of OSN data is massive and cannot be handled efficiently by traditional streaming and time series clustering algorithms [31]. In order to tackle that problem, we propose a novel approach to speed up the processing of online meme clustering that uses both semantic similarity and Wikipedia concepts to efficiently store and summarize OSN data in real time.

With the increasing amount of online social network data, understanding and analyzing them is becoming more challenging. Researchers have started to employ human's ability to effectively gain visual insight on data analysis tasks. The task of clustering memes shares some similarity with clustering text, but they are also intrinsically different. For example, social network data is usually poorly written and content-limited. This reduces the quality of clustering results. For a Reddit submission, the relationships between the title, comment, image, and URL sometimes are disconnected (e.g., a title has a different subject from the content). In this paper, we developed a visualization prototype to allow users to better distinguish the similarity between submissions and use this feedback to improve the clustering results.

This paper extends previous work of Dang et al. [18] by formalizing the problem of meme clustering and proposes a novel approach for clustering Reddit submissions. It makes the following contributions:

- Extends and improves the similarity scores between different elements of Reddit submission of Dang et al. [18] by introducing the use of Wikipedia concepts as an external knowledge.
- Introduces a modified version of Jaccard Coefficient that employs the use of text semantic similarity when comparing the similarity score between two sets of Wikipedia concepts.
- Proposes an Offline–Online clustering algorithm that exploits semantic similarity and Wikipedia concepts to achieve good clustering results in real time. The offline clustering component computes and summarizes cluster statistics to speed up the process of the online clustering component. In addition, for each cluster, we adopt the damped window model and propose a novel approach to summarize each cluster as a set of Wikipedia concepts where each concept is assigned a weight based on its recency and popularity. The online clustering component applies a semantic version of Jaccard Coefficient.
- The experiments show the use of Wikipedia concepts increases the accuracy result of the meme clustering tasks. Although only using Wikipedia concepts as a similarity score does not increase the clustering result, using both Wikipedia concepts and text semantic similarity increase the clustering accuracy for both offline and online clustering components.

2   Related Work

This section presents current research on text semantic similarity and detecting the spread of memes in OSNs.

2.1   Similarity Measures and Text Clustering

Several similarity measures have been proposed in the literature for the task of text clustering. The most popular ones are lexical measures like Euclidean, Cosine, Pearson Correlation, and Extended Jaccard measures. Strehl et al. [40] provided a comprehensive study on using different clustering algorithms with these people measures. The authors used several clustering algorithms on the YAHOO dataset, and showed that Extended Jaccard and Cosine similarity performed better and achieved results that are close to a human-labelling process. However, lexical similarity measures do not consider the semantic similarity between words in the texts.

Some researchers have taken advantage of the semantic relatedness of texts by using external resources to enrich word representation. In [38], the authors suggested using WordNet as a knowledge base to determine the semantic similarity between words. The experiment results have shown that external knowledge bases like WordNet improve the clustering results in comparison to the Bag-of-Words models. Hu et al. [28] proposed the use of Wikipedia as external knowledge for text clustering. The authors tried to match concepts in text into Wikipedia concepts and categories. Similarity scores between concepts are calculated based on the text content information, as well as Wikipedia concepts, and categories. The experiment results have shown that using Wikipedia as external knowledge provided a better result than using WordNet due to the limited coverage of WordNet. Bollegala et al. [8] proposed the use of information available on the Web to compute text semantic similarity by exploiting page counts and text snippets returned by a search engine. Our work is intuitively different from these approaches, as it introduces the use of word relatedness based on the Google n-grams dataset [9]. The proposed semantic similarity scores between texts are calculated based on that algorithm to handle the low quality (i.e., poor writing) of social network data. Using Google n-grams dataset as external knowledge is more effective than textual as well as other semantic approaches, as the Google n-grams dataset has more coverage than other semantic approaches.

2.2 Online Clustering Algorithms in OSNs

This section discusses the related work of event detection and online meme clustering in OSNs. The proposed online meme clustering algorithm takes advantage of the current work of both clustering streaming data and clustering time series data. Clustering streaming data has been actively researched in the literature. Aggarwal et al. [3] proposed a graph-based sketch structure to maintain a large number of edges and nodes at the cost of potential loss of accuracy. Zhao and Yu [42] extended the graph-based clustering streaming algorithms with side information, such as user metadata in OSNs. As OSN data is dependent on its temporal context, time series is another important feature of clustering streaming data algorithms. We present related work of the three types of time series clustering algorithms in the literature: (1) landmark window approach, (2) sliding window approach, and (3) damped window approach [39]. In clustering streaming data, landmark-based models consider all the historical data from a landmark time and all data have an equal weight [11, 24, 36]. Sliding-based models are common stream processing models which only examine data at a fixed-time window (e.g., last 5 min or last 24 h) [14, 25, 35]. Damped window models introduced the use of decay variable to replace old data to increase the accuracy of streaming clustering results [13, 23]. JafariAsbagh et al. [31] used a sliding window approach for detecting memes in real time that does not consider the topic evolution and persistence. As the spread of memes in OSNs is dependent on the meme topics and its context [26], the proposed online meme clustering algorithm explores the damped window approach which considers the frequency and recency of memes. Researchers also investigate if the use of external knowledge (e.g., Wikipedia) helps the clustering results for social media texts. Banerjee et al. [4] introduced the use of Wikipedia as external knowledge to improve the accuracy results for short texts. Dang et al. [18] used text semantic similarity computed from Google n-grams dataset to alleviate the problem of shortness and noise of OSN data.

Scientists also explored the use of visualization for text clustering with relevance user feedback. Lee et al. [33] introduced iVisClustering, an interactive visualization framework based on LDA topic modelling. This system provides some interactive features, such as removing documents or clusters, moving a document from one cluster to another, merging two clusters, and influencing term weights. Choo et al. [16] presented an interactive visualization for dimension reduction and clustering for large-scale high-dimensional data. The system allows users to interactively try different dimension reduction techniques and clustering algorithms to optimize the clustering results. One of the limitations of these systems is that they focus on the clustering algorithms and results and have limited supports for combining similarity scores for different parts of a text (e.g., the title and body of a text). This paper introduces a visualization prototype to combine different similarity scores for our clustering process interactively and incrementally.

2.3 Detecting Memes in Online Social Networks

Recently, researchers have started adapting state-of-the-art clustering algorithms to OSN data. Leskovec et al. [34] proposed a meme-tracking framework to monitor memes that travel through the Web in real time. The framework studied the signature path and topic of each meme by grouping similar short, distinctive phrases together. One drawback of this framework is that it only applies lexical content similarity to detect memes. This did not work well for memes that are related but not using the same words, and those that are short and concise (e.g., Tweets on Twitter). Cataldi et al. [10] proposed an approach that monitored the real-time spread of emerging memes in Twitter. The authors defined an emerging term as one whose frequency of appearance had risen within a short period and had not emerged or was only rarely discussed in the past. A navigable topic graph is constructed to connect semantically related emerging terms. Emerging memes are extracted from this graph based on semantic relationships between terms over a specified time interval. Becker et al. [5] formulated the problem of clustering for event detection and proposed a supervised approach to classify tweets using a predefined set of features. The proposed approach includes various types of features: textual, temporal, and spatial. Aggarwal and Subbian [1] presented a clustering algorithm that exploits both content and network-based features to detect events in social streams. The proposed algorithm uses knowledge about metadata of Twitter users. Thom et al. [41] developed a system for interactive analysis of location-based microblog messages, which can assist in the detection of real-world events in real time. This approach uses X-means, a modified version of K-means, to detect emerging events. Finally, JafariAsbagh et al. [31] introduced an online meme clustering framework using the concept of Protomemes. Each Protomeme is defined based on one of the atomic information entities in Twitter: hashtags, mentions, URLs, and tweet content. An example of Protomeme is the set of tweets containing the hashtag #All4Given. This approach uses a sliding window model that can lead to good offline prediction accuracy but not suitable for online streaming environments. As online meme clustering algorithms require low prediction and training costs, our proposed online meme clustering algorithm stores cluster summary statistics using Wikipedia concepts and applies a damped window approach with Offline–Online components for clustering memes in OSNs. Although Twitter has been the most popular OSN for detecting memes, little work has been done to detect rumour-related memes on Reddit.

3 Reddit Social Network

Reddit, which claims to be "the front page of the internet," is a social news website, where users, called redditors, can create a submission or post direct links to other online content. Other redditors can comment or vote to decide the rank of this submission on the site. Reddit has many subcategories, called sub-reddits that are organized by areas of interests. The site has a large base of users who discuss a wide range of topics daily, such as politics and world events. Alexa ranks Reddit.com as the 24th most visited site globally. Each Reddit submission has the following elements:

- **Title:** The title summarizes the topic of that submission. The title text is usually very short and concise. The title may also have a description to further explain it.
- **Comments:** Users can post a comment that expresses their opinions about the corresponding submission or other user comments. Users can also vote comments up or down.
- **URL:** Each submission may contain a link to an external source of information (e.g., news articles) that is related to the submission.
- **Image:** Submissions may also have a link to an image that illustrates the topic of the submission.

Figure 3 explains how to collect image and URL content from Reddit submissions. Unlike other OSNs, Reddit is fundamentally different in that it implements an open data policy; users can query any posted data on the website. For example, other

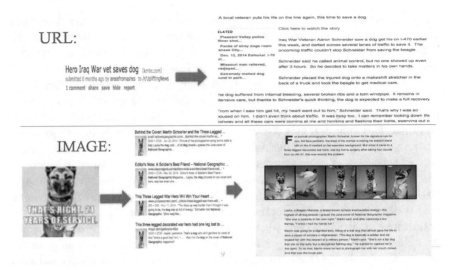

Fig. 3 External content from image and URL. The *top* submission has a URL and we extracted the URL content. The *bottom* submission has an image and we extracted the text of the image from Google Reverse Image Search

OSNs, like Twitter or Facebook, allow circulating information through a known cycle (e.g., "follow" connections), whereas Reddit promotes a stream of links to all users in a simple bookmarking interface. This makes Reddit a effective resource to study the spread of memes in OSNs [19, 20]. To the best of our knowledge, no similar work has been done on clustering memes in Reddit.

4 Google Tri-gram Method

Google Tri-gram Method (GTM) [30] is an unsupervised corpus-based approach for computing semantic relatedness between texts. GTM uses the uni-grams and tri-grams of the Google Web 1T N-grams corpus [30] to calculate the relatedness between words, and then extends that to longer texts. The Google Web 1T N-grams corpus contains the frequency count of English word n-grams (unigrams to 5-g) computed over one trillion words from web page texts collected by Google in 2006.

The relatedness between two words is computed by considering the tri-grams that start and end with the given pair of words, normalizing their mean frequency with unigram the frequency of each of the words as well as the most frequent unigram in the corpus as shown in Fig. 4, where $C(\omega)$ is the frequency of the word ω. $\mu_T(\omega_1, \omega_2)$ is the mean frequency of trigrams that either start with ω_1 and end with ω_2, or start with ω_2 and end with ω_1. $\sigma(a_1, \ldots, a_n)$ is the standard deviation of numbers $a_1, \ldots, a_n$, and $C_{\max}$ is the maximum frequency among all unigrams.

$$GTM(\omega_1, \omega_2) = \begin{cases} \dfrac{\log \frac{\mu_T(\omega_1,\omega_2)C_{max}^2}{C(\omega_1)C(\omega_2)\min(C(\omega_1)C(\omega_2))}}{-2\times\log\frac{\min(C(\omega_1),C(\omega_2))}{C_{max}}} & \text{if } \log\frac{\mu_T(\omega_1,\omega_2)C_{max}^2}{C(\omega_1)C(\omega_2)\min(C(\omega_1)C(\omega_2))} > 1 \\[4mm] \dfrac{\log 1.01}{-2\times\log\frac{\min(C(\omega_1),C(\omega_2))}{C_{max}}} & \text{if } \log\frac{\mu_T(\omega_1,\omega_2)C_{max}^2}{C(\omega_1)C(\omega_2)\min(C(\omega_1)C(\omega_2))} \leq 1 \\[4mm] 0 & \text{if } \mu_T(\omega_1,\omega_2) = 0 \end{cases}$$

Fig. 4 GTM semantic similarity calculation [30]

GTM computes a score between 0 and 1 to indicate the relatedness between two texts based on the relatedness of their word content. For given texts P and R where $|P| \leq |R|$, first all the matching words are removed, and then a matrix with the remaining words $P' = \{p_1, p_2, \ldots, p_m\}$ and $R' = \{r_1, r_2, \ldots, r_n\}$ is constructed where each entry is a GTM word relatedness $a_{ij} \leftarrow GTM(p_i, r_j)$.

$$M = \begin{pmatrix} a_{11} & a_{12} & \cdots & a_{1n} \\ a_{21} & a_{22} & \cdots & a_{2n} \\ \vdots & \vdots & \ddots & \vdots \\ a_{m1} & a_{m2} & \cdots & a_{mn} \end{pmatrix}$$

From each row $M_i = \{a_{i1} \cdots a_{in}\}$ in the matrix, significant elements are selected if their similarity is higher than the mean and standard deviation of words in that row:

$$A_i = \{a_{ij} | a_{ij} > \mu(M_i) + \sigma(M_i)\},$$

where $\mu(M_i)$ and $\sigma(M_i)$ are the mean and standard deviation of row i. Then the document relatedness can be computed using:

$$Rel(P, R) = \frac{(\delta + \sum^m a_{i=1}\sigma(A_i)) \times (m+n)}{2mn}$$

where $\sum^m a_{i=1}\sigma(A_i)$ is the sum of the means of all the rows, and δ is the number of removed words when generating P' or R'.

5 Semantic Jaccard Coefficient

Jaccard similarity coefficient is a statistic used to compute the similarity and diversity between two sets. Chierichetti et al. [15] showed that finding an optimal solution for weighted Jaccard median is an NP-hard problem and presented a heuristic algorithm to speed up the computational complexity. The Jaccard coefficient between two sets A and B is defined as follows:

$$J(A, B) = \frac{A \cap B}{A \cup B} \text{ where } 0 \leq J(A, B) \leq 1$$

We propose a modified version of Jaccard coefficient that exploits the use of semantic similarity using GTM. As the original Jaccard coefficient only uses an exact pattern matching, it does not work well if two Wikipedia concepts are not the same but are semantically similar. For example, the Jaccard coefficient for two concepts "President of the United States" and "Barack Obama" should be high as they are semantically similar using GTM.

For two submissions $S_1 = \{T_{11}, T_{12}, \ldots, T_{1n}\}$ and $S_2 = \{T_{21}, T_{22}, \ldots, T_{2n}\}$ where T_i is a Wikipedia concept extracted from the title or comments of submission S_i, the Semantic Jaccard Coefficient (SJC) is defined as:

$$SJC(S_1, S_2) = \frac{S_1 \cap S_2}{S_1 \cup S_2} \text{ where } 0 \leq SJC(S_1, S_2) \leq 1 \tag{1}$$

where T_{1i} and T_{2j} are semantically equivalent, $T_{1i} \equiv T_{2j}$, if $GTM(T_{1i}, T_{2j}) \geq e$, where e is a parameter that is explored through the experiment. If T_i is semantically similar to more than one concept in S_2, we use the concept with the highest GTM score.

6   Similarity Scores and Combination Strategies

This section explores the use of GTM semantic similarity of Dang et al. [18] and introduces Wikipedia concepts as an external knowledge to propose five semantic similarity scores and their combinations between submissions. Representing a submission S in Reddit as a vector $S = (T, M, I, U, W)$ where:

- T is an n-dimensional feature vector $t_1, t_2, \ldots t_n$ representing the title of the submission and its description.
- M is an n-dimensional feature vector $m_1, m_2, \ldots m_n$ representing the comments of a submission.
- U is an optional n-dimensional feature vector $u_1, u_n, \ldots u_n$ representing the external URL content of a submission.
- I is an optional n-dimensional feature vector $i_1, i_2, \ldots i_n$ representing the image content of a submission. This content is extracted by using Google reverse image search, which takes an image as a query and extracts the text content of the website that is returned from the top search result and is not from Reddit.
- W is an optional n-dimensional feature vector $w_1, w_2, \ldots w_n$ representing the Wikipedia concepts of the titles and comments of a submission.

6.1   Similarity Scores

We propose five similarity measures between two submissions S_1 and S_2:

- **Title similarity** SC_t is the GTM semantic similarity score between the title word vectors T_1 and T_2.
- **Comment similarity** SC_m is the GTM semantic similarity score between the comment word vectors M_1 and M_2.
- **URL similarity** SC_u is the GTM semantic similarity score between the URL content word vectors U_1 and U_2.
- **Image similarity** SC_i is the GTM semantic similarity score between the word vectors I_1 and I_2 retrieved from Google Reverse Image Search.
- **Wikipedia similarity** SC_w is the SJC score between the bag of concept vectors W_1 and W_2 retrieved from titles and comments of submissions using Eq. (1).

6.2   Combination Strategies

The main goal of this section is to study the effect of different similarity scores and their combinations on the quality of the meme clustering tasks. We incorporate Wikipedia concepts as an external knowledge to all combination strategies from our previous work [18].

6.2.1   Pairwise Maximization Strategy

The pairwise maximization strategy chooses the highest among the title, comment, URL, and image scores to decide the similarity between two submissions. This strategy avoids the situation where similarity scores have a low content quality (e.g., titles are short and lack details, comments are noisy, images and URLs are not always available) by choosing the most similar among them.

Given two submissions $S_1 = \{T_1, M_1, I_1, U_1, W_1\}$ and $S_2 = \{T_2, M_2, I_2, U_2, W_2\}$, the pairwise maximization strategy between them is defined as:

$$\text{MAX}_{S_1 S_2} = \text{MAX}(\text{GTM}_{T_1 T_2}, \text{GTM}_{M_1 M_2}, \text{GTM}_{U_1 U_2}, \text{GTM}_{I_1 I_2}, \text{SJC}_{W_1 W_2}) \quad (2)$$

where $\text{GTM}_{T_1 T_2}, \text{GTM}_{M_1 M_2}, \text{GTM}_{U_1 U_2}, \text{GTM}_{I_1 I_2}$ are the title, comment, URL, and image similarity scores between the two submissions S_1 and S_2. $\text{SJC}_{W_1 W_2}$ is the SJC score between two submission S_1 and S_2 using Eq. (1) for the Wikipedia concepts extracted from submission titles and comments.

6.2.2 Pairwise Average Strategy

The pairwise average strategy computes the average value of the five pairwise similarity scores. This strategy balances the scores among the five similarities in case some scores do not reflect the true content of the submission. It is defined as follows:

$$\text{AVG}_{S_1 S_2} = \text{AVG}(\text{GTM}_{T_1 T_2}, \text{GTM}_{M_1 M_2}, \text{GTM}_{U_1 U_2}, \text{GTM}_{I_1 I_2}, \text{SJC}_{W_1 W_2}) \quad (3)$$

6.2.3 Linear Combination Strategy

In the linear combination strategy, users can assign different weighting values manually. For example, if users think the title text does not capture the topic of a submission, they can assign a low weight factor (e.g., 0.1). If they think comment texts are longer and represent the topic better, they can assign a higher weight factor (e.g., 0.6). The linear combination strategy is defined as follows:

$$\text{LINEAR}_{S_1 S_2} = \text{LINEAR}(w_t \text{GTM}_{T_1 T_2}, w_m \text{GTM}_{M_1 M_2}, w_u \text{GTM}_{U_1 U_2}, w_i \text{GTM}_{I_1 I_2}, w_w \text{SJC}_{W_1 W_2}) \quad (4)$$

where w_t, w_m, w_u, w_i, and w_w are the weighting factors for titles, comments, images, urls, and Wikipedia concepts with a normalization constraint $w_t + w_m + w_u + w_i + w_w = 1$.

6.2.4 Internal Centrality-Based Weighting

Computing the optimized weight factors for the linear combination strategy is an intensive task. JafariAsbagh et al. [31] used a greedy optimization algorithm to compute the optimized linear combination for the task of clustering memes. However, it is unrealistic to compute all the possible weighting combinations for Eq. (4). To alleviate this computational cost, we propose the Internal Centrality-Based Weighting (ICW), a novel strategy to automatically calculate the weight factors of the linear combination strategy. This strategy calculates the weight factors for each element of a submission by considering its surrounding context. Although all elements of a submission are semantically related, some elements could have more semantic content than others; for example, the URL content discusses more the topic than the title. More weight is assigned to the elements with higher semantic content. The proposed strategy is shown in Eq. (5). It computes the semantic content weights using internal and external similarity scores between titles, comments, URLs, images, and Wikipedia concepts of two submissions. We append all the Wikipedia concepts together to compute the GTM score between Wikipedia concepts and other texts. For each submission, this strategy computes the centrality score for each element of each submission S_i:

$$\text{CENT}_{T_i} = \text{GTM}_{TM_i} + \text{GTM}_{TU_i} + \text{GTM}_{TI_i} + \text{GTM}_{TW_i}$$

$$\text{CENT}_{M_i} = \text{GTM}_{MT_i} + \text{GTM}_{MU_i} + \text{GTM}_{MI_i} + \text{GTM}_{MW_i}$$

$$\text{CENT}_{U_i} = \text{GTM}_{UT_i} + \text{GTM}_{UM_i} + \text{GTM}_{UI_i} + \text{GTM}_{UW_i}$$

$$\text{CENT}_{I_i} = \text{GTM}_{IT_i} + \text{GTM}_{IM_i} + \text{GTM}_{IU_i} + \text{GTM}_{IW_i}$$

$$\text{CENT}_{I_w} = \text{GTM}_{WT_i} + \text{GTM}_{WM_i} + \text{GTM}_{WU_i} + \text{GTM}_{WI_i}$$

Then, it computes the weighting factors between two submissions S_1 and S_2 by:

$$w_T = \text{CENT}_{T_1} * \text{CENT}_{T_2}$$

$$w_M = \text{CENT}_{M_1} * \text{CENT}_{M_2}$$

$$w_U = \text{CENT}_{U_1} * \text{CENT}_{U_2}$$

$$w_I = \text{CENT}_{I_1} * \text{CENT}_{I_2}$$

$$w_W = \text{CENT}_{W_1} * \text{CENT}_{W_2}$$

Then, it normalizes the weighting factors so that: $w_T + w_M + w_U + w_I + w_W = 1$, and finally computes the ICW strategy:

$$\text{ICW}_{S_1 S_2} = \text{ICW}(w_T \text{GTM}_{T_1 T_2}, w_M \text{GTM}_{M_1 M_2}, w_U \text{GTM}_{U_1 U_2}, w_I \text{GTM}_{I_1 I_2}, w_W \text{GTM}_{W_1 W_2}) \quad (5)$$

6.2.5 Similarity Score Reweighting with Relevance User Feedback

One effective way to improve the clustering results is to manually specify the relationships between pairwise documents (e.g., must-link and cannot-link) to guide the document clustering process [7]. As social network data are intrinsically heterogeneous and multidimensional, it is not easy to compare two submissions to determine if they are similar or not without putting them into the same context. To overcome this limitation, a novel technique, the Similarity Score Reweighting with Relevance User Feedback (SSR), is proposed to incorporate relevance user feedback by a visualization prototype in which submissions are displayed as a force-directed layout graph where:

- **A node** is a submission in Reddit.
- **An edge** is a connection between two submissions if their similarity scores are above a threshold (default 0.85).
- **A node color** represents to which cluster it belongs.

Algorithm 1 describes how the visualization system integrates user feedback to remove outliers, move submissions from a cluster to another, or reassign similarity score weighting factors for submissions. Users can select any of the five proposed

Algorithm 1 Semi-supervised similarity score reweighting with relevance user feedback strategy (SSR)

Input: a set of submissions X from Reddit.
Output: K clusters $\{X\}_{l=1}^{K}$
1: **loop**
2:     {**Step 1**} Perform k-means clustering on P percent of the ground-truth dataset using one of the proposed strategies. P is defined through experiments.
3:     {**Step 2**} Visualize the clustering result in step 1.
4:     {**Step 3**} Allow users to interactively remove outlier submissions, reassign submission class labels, or assign weight factors for each element between two submissions.
5:     {**Step 4**} Re-cluster the submissions based on user inputs.
6:     {**Step 5**} repeat step 1 if necessary.
7: **end loop**
8: {**Step 6**} Recluster the whole dataset considering user feedback in Step 1 to 5.

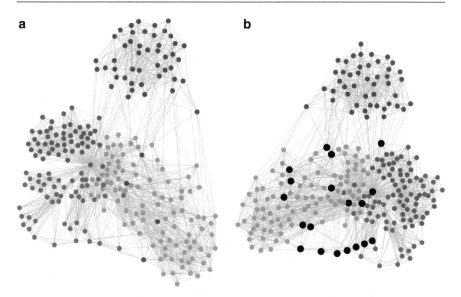

Fig. 5 The proposed meme visualization: (**a**) The original visualization graph and (**b**) The clustering result using ICW

strategies, MAX, AVG, LINEAR, and ICW as a baseline for clustering. Figure 5a shows an SSR visualization of the meme dataset using ICW strategy. The graph has five different colors that represent five memes in the ground-truth dataset. Users can pan, zoom, or click on a submission to get more details about this submission. They can also click on the checkbox "Show wrong cluster assignments" to see which submissions are incorrectly assigned by the ICW strategy. Based on the graph visualization, users can understand how a submission is positioned regarding its neighbor submissions. When clicking on a node in the graph, users will be redirected to the actual submission in Reddit to find out more information and decide if it belongs to the correct cluster. Most of the incorrectly clustered are overlapped or outlier nodes as shown in Fig. 5b. For each incorrectly assigned submission, users can remove, update its class label, or assign a different similarity coefficient score for each element between two submissions. SSR focuses on human knowledge to detect outliers or borderline submissions.

7   The Offline–Online Meme Detection Framework

The meme detection problem is defined for any social media platform used to spread information. In these systems, users can post a discussion or discuss a current submission. An overview of the proposed meme detection framework is shown in Fig. 6.

7.1   The Offline–Online Meme Clustering Algorithm

As OSN data is changing and updating frequently, we modify and extend the proposed ICW algorithm [18] to work with the online streaming clustering algorithm using the semantic similarity and Wikipedia concepts to handle continuously evolving data over time. As emerging events or topics are changing in real time, some topics may appear but not burst. Other topics or events may appear and become a popular topic for a long period. Based on this observation, the proposed framework adopts the damped window model [23] and assigns more weight to recent data and popular topics. It also adopts the Offline–Online components of Aggarwal et al. [2] to make the online meme clustering more efficient. As clustering OSN data is a computationally intensive task, the offline component does a one-pass clustering for existing OSN data in the first step. It also calculates and summarizes each cluster statistics using Wikipedia concepts extracted from the titles and comments of all

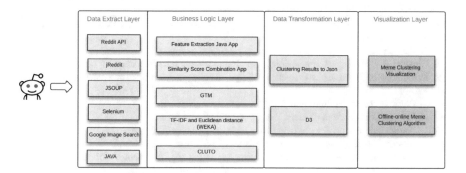

Fig. 6 The proposed meme detection framework

the submissions in the same cluster. For the online component, it assigns new data points to the appropriate clusters using a modified version of online k-means.

For an online stream $S_1, S_2, \ldots, S_n$ where each S_i is a submission in Reddit. Each submission S_i is represented by a 5-tuple (T, C, U, I, W) that represents the title, comments, URL, image, and Wikipedia concepts (of titles and comments) of this submission. At a time t, the proposed framework are presented in two steps:

- **Offline component:** cluster all the submission from $t - 1$ to t_0 into k cluster $C_1, C_2, \ldots, C_n$, such that.

  - Each submission S_i belongs to only one cluster.
  - The submissions are clustered into clusters using the ICW strategy.

- **Online component:** assign an incoming submission S_i into one of the clusters created from the offline component.

The online k-means and the sliding window model of JafariAsbagh et al. [31] do not consider the popularity and occurrence frequency of a topic. To overcome this problem, we proposed an approach to compute the popularity of topics and use it as a parameter for the damped window model. Each cluster is represented by a set of Wikipedia concepts $w_1, w_2, \ldots w_n$ and each concept can be linked back to the original submissions. Each concept in the cluster statistics is represented by the exponential decay function:

$$W(t) = N * e^{-\lambda t} \tag{6}$$

where N is the count of this concept in the cluster. λ is a positive exponential decay constant. If a concept stays in the cluster for a period but there are no new submissions that contain this concept, it will be removed from the cluster set. A summary of the proposed algorithm is shown in Algorithm 2 and its explanation is shown in Fig. 7.

Algorithm 2 The Offline–Online meme clustering algorithm

Input: a set of submissions $S = \{S_1, S_2, \ldots, S_n\}$ from Reddit at time t. k = number of clusters.
    m = number of concepts in each cluster.
 1: Collect all submissions in Reddit for an intial period of time t.
 2: Cluster all the collected submissions using ICW strategy into k clusters $C_0, C_1, \ldots, C_k$.
 3: Summarize each cluster by extracting Wikipedia concepts from titles and comments.
 4: **loop**
 5:    Retrieve the next submission S_{t+1}.
 6:    For each summarized cluster C_i, compute SJC(S_i, C_i) score.
 7:    Assign submission S_{t+1} to the cluster with the highest SJC score.
 8:    Re-compute the cluster summary statistics of the selected cluster.
 9:    Only keeps m concepts in each cluster using Eq. (6). If a submission has no existing concepts in the summary, remove this submission from the cluster.
10: **end loop**

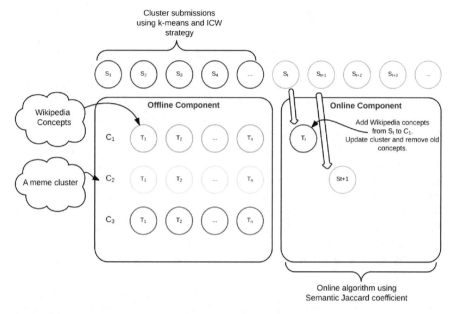

Fig. 7 The Offline–Online meme clustering algorithm

For the online component, when a new submission is assigned to a cluster, it may not naturally belong to this cluster. For example, this submission can be an outlier or the beginning of a new cluster. Equation (1) handles these two situations naturally. If the submission is an outlier, it will be removed afterward if there are no similar incoming submissions that contain concepts from this submission. For the second situation, the cluster will be naturally replaced by the new concepts and the old concepts will be removed.

8 Experimental Results

The objective of this section is to evaluate the performance of the meme clustering tasks with the incorporation of Wikipedia concepts and the proposed online meme clustering algorithms. We re-evaluate the first three experiments from our previous work [18] with the incorporation of Wikipedia concepts and further access three new experiments. First, we explain how the ground-truth dataset is extracted from Reddit, and then discuss the evaluation metric and how the experiments are carried out.

8.1   Ground-Truth Dataset

To study the spread of memes in Reddit, the posts and comments related to a specific meme are identified. A generic query is used to capture all of the related submissions for a specific meme. Since there are no available Reddit meme datasets, RedditAPI and jReddit, an open source Java project, are used to extract submissions, comments, and other data views (image and URL content) about a specific meme using predefined regular expressions. All submissions that do not have any comments or "up" or "down" votes are removed, as we assume that users are not interested in them. In addition, comments less than 5 words long are ignored. Stop words are also removed. For each submission with a URL in its title, JSOUP is used to parse the main body text content of the URL. Occasionally, a submission can have an image in its title. Selenium is used to submit the image to Google Reverse Image Search to find the most similar webpage to this image. If the top-searched result returns an article from Reddit, the program traverses through the search result list until it finds an article that is not from Reddit. Wikipedia concepts are extracted using Dexter [12].

The ultimate goal of this framework is to detect memes and discussion topics online. In order to access the performance of the proposed similarity strategies, we collect ground-truth data for the experiments. First, the five most popular topics in Reddit from October to November 2014 are selected. The program extracts titles, comments, URL, and image content of all related submissions for each topic. Each topic is labeled to the corresponding cluster based on the keyword search. The five topics (clusters) are: (1) EBOLA (2) Ferguson (3) ISIS (4) Obama and (5) Trayvon Martin. Table 1 shows the detailed statistics of the ground-truth dataset.

8.2   Clustering Algorithms

The paper adopts k-means clustering as the baseline-clustering algorithm because of its simplicity and efficiency. The experiments focused more on determining if the proposed similarity strategies improve clustering results and the proposed online clustering algorithm could detect emerging memes in real time. GTM is used to compute the similarity score between texts. The output of the GTM algorithm is a similarity matrix that shows the similarity score for each text with the other texts in the dataset. After producing this similarity matrix, gCLUTO is used to cluster the matrix using an equivalent version of k-means clustering.

Table 1 The experiment ground-truth dataset

No.	Topic	Submission counts	Comments	Submissions with a URL	Submissions with an image
1	EBOLA	495	89,394	218	39
2	FERGUSON	495	83,912	203	87
3	ISIS	488	76,375	190	61
4	OBAMA	490	139,478	142	13
5	Trayvon Martin	471	93,848	250	30

8.3 Baselines

For the baselines, each title, comment, URL, or image text is represented as a TF-IDF vector. Euclidean distance [21] is used to calculate the similarity score between TF-IDF vectors due to its simplicity. In the next section, we compare the clustering results of the proposed strategies and algorithms with the baseline.

8.4 Results

For the ground-truth dataset, since the class labels exist for all of the submissions, purity is adopted (i.e., the number of correctly assigned submissions over the total number of submissions) as an evaluation measure. Larger purity value indicates better clustering results. Several configurations are explored to evaluate the performance of the proposed similarity strategies. As URL, image content, and Wikipedia concepts are not always available, they are used as additional data for the clustering tasks for MAX, AVG, and ICW. The proposed configurations are used for both GTM similarity and baseline similarity and configured as:

- **TITLE**: Only use the title similarity for pairwise submission comparison.
- **COMMENT**: Only use the comment similarity for pairwise submission comparison.
- **MAX**: Use the maximum of the five similarity scores for pairwise submission comparison as defined in Eq. (2).
- **AVG**: Use the average of the five similarity scores for pairwise submission comparison as defined in Eq. (3).
- **ICW**: Calculate the pairwise similarity between two submissions based on internal centrality weighting as defined in Eq. (5).

8.4.1 Clustering with Semantic Similarity Scores

The first experiment explores the advantage of using GTM semantic similarity and Wikipedia concepts for meme clustering tasks. The k-means clustering results between the proposed similarity scores and the baselines using TF-IDF and Euclidean are compared for TITLE, COMMENT, MAX, and AVG. GTM score outperforms the baselines as shown in Figs. 8 and 9. Using comment content of a submission for the meme clustering task produces a better result than using title content as title texts are usually concise and does not represent the context of a submission. Exploiting additional image and URL content by AVG and MAX strategies improves the clustering results as shown in Figs. 10 and 11. Another interesting result is that the performance of GTM for comments is very close to

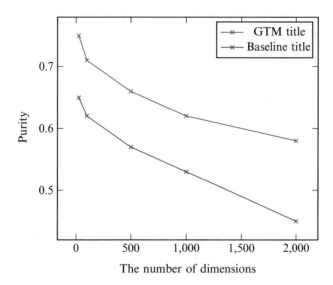

Fig. 8 GTM title vs. Baseline title

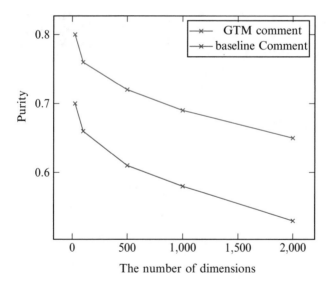

Fig. 9 GTM comment vs. Baseline comment

the AVG strategy. Using AVG strategy does not capture the semantic content of each similarity score efficiently. The experiment results also show that GTM & Wikipedia score scales better than the baseline for higher vector dimensions. We conjecture that GTM & Wikipedia help alleviate "the curse of dimensionality" for clustering using traditional similarity measures.

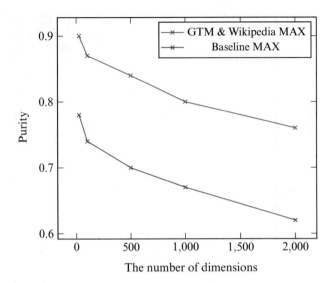

Fig. 10 GTM & Wikiepdia MAX vs. Baseline MAX

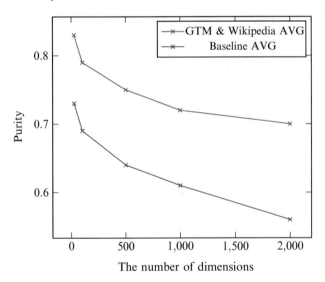

Fig. 11 GTM & Wikipedia AVG vs. Baseline AVG

8.4.2 The ICW Strategy

In this experiment, the objective is to find out if the proposed ICW strategy with the use of GTM and Wikipedia concepts improves the clustering result for a meme clustering task. The experiment results between the proposed ICW, AVG, and MAX are shown in Fig. 12. Results indicated that ICW outperforms AVG and achieves

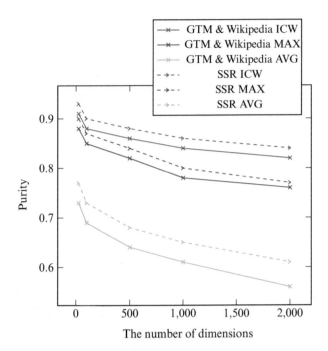

Fig. 12 Clustering results with different similarity score strategies

better results than MAX. We also found that the AVG combination does not provide good results when comparing with using MAX or ICW. One of the reasons may be each similarity score plays a different role in distinguishing memes in Reddit and this agrees with our assumption about the semantic content related between elements of submissions.

8.4.3 Similarity Score Reweighting with Relevance User Feedback

This section investigates the improvement from using user feedback with the visualization prototype for the meme clustering task. In the first step, users choose one of the five proposed strategies (AVG, MAX, LINEAR and ICW) to cluster the ground-truth dataset. For the LINEAR, we explore different weight factors for the Eq. (4). Although the clustering results are improved when weight factors for title and comment are low (e.g., 0.1 for titles, 0.3 for comments) and are high for URLs and image (e.g., 0.6), their results are still not optimized when comparing with MAX and ICW. We remove outliers and reassign the weight factors for overlapping nodes. The clustering results are statistically improved for both MAX, AVG, and ICW at $p = 0.05$ as shown in Fig. 12 (SSR ICW, SSR MAX, SSR AVG).

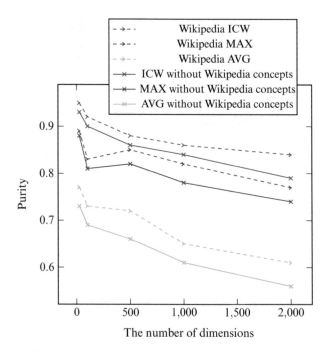

Fig. 13 Clustering results with Wikipedia concepts

8.4.4 Similarity Score with Wikipedia Concepts

This section studies whether the use of Wikipedia concepts improves the clustering results. We compared the clustering results with the three combination strategies (MAX, AVG, and ICW). We also explore whether the use of titles and Wikipedia concepts alone produces a good clustering result. The clustering results are statistically improved for MAX, AVG, and ICW at $p = 0.05$ as shown in Fig. 13. We also observed that using only titles and Wikipedia concepts did not achieve good clustering results. One of the underlying reasons is because there are overlapping concepts between memes which downgrades the clustering results.

8.4.5 Semantic Jaccard Coefficient

This section inspects whether the proposed Jaccard coefficient helps to increase the accuracy of the meme clustering results. We investigate whether the semantic Jaccard coefficient increases the clustering results of MAX and AVG ($e = 0.8$). We find that Jaccard coefficient using semantic similarity significantly increases the clustering results, as shown in Fig. 14. This happens because there are many Wikipedia concepts that are very similar but not the same in the ground-truth dataset,

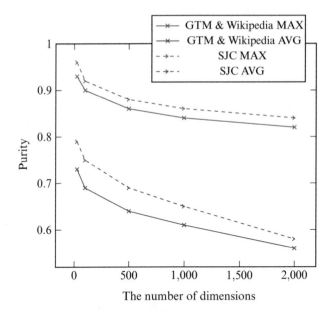

Fig. 14 The proposed modified Jaccard coefficient vs. MAX, AVG

for example, "Obama" vs. "President." Combining the Wikipedia concepts with the semantic similarity scores using SJC significantly improves the clustering results.

8.4.6 The Offline–Online Meme Clustering Algorithm

This section examines whether the proposed Offline–Online clustering algorithm achieves a good clustering result. We use the ground-truth dataset to evaluate the algorithm. For the offline component, we select the first 2000 submissions from the ground-truth and cluster them into five clusters using the ICW strategy. For each cluster, we extract all the Wikipedia concepts from titles and comments of each submission as the cluster summary. We process the remaining 500 submissions using the online component in an ascending time order. Each incoming submission is assigned to its closest cluster using Eq. (6) ($\lambda = 1$). Finally, the clustering accuracy results between the Offline–Online clustering algorithms and ICW are shown in Fig. 15 where the x-axis represents the increasing number of concepts. The experiment results show that although the proposed online clustering algorithm did not perform as well as the ICW with a low number of concepts, the results are comparable when the number of concepts is higher than 10,000. In addition, although using only Wikipedia concepts does not work well for the offline mode, it achieves a reasonable accuracy for the Offline–Online algorithm. The underlying reason is that we use the offline component as a learning model using ICW strategy, and it helps alleviate the noisy nature of OSN data.

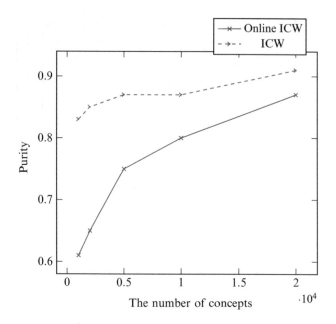

Fig. 15 The proposed online clustering algorithms vs. ICW

9 Conclusions

This paper presents an online framework to tackle the problem of meme clustering in Reddit as a means to detect emerging events and rumour-driven topics and their spread as specific clusters. This framework makes use of Google Web 1T n-gram corpus (GTM algorithm) to compute the similarity between texts and Wikipedia concepts as external knowledge for the meme clustering task. It also defines several pairwise similarity scores between elements of two submissions. These scores include external content related to image and URL elements of a submission. The paper explores a semantic similarity version of Jaccard coefficient and several strategies to combine the similarity scores in order to produce better clustering results. These strategies include average, maximum, linear combination, internal centrality-based weighting, and similarity score reweighting with relevance user feedback. Finally, it proposes an Offline–Online meme clustering framework to both detect memes in real time and achieve good clustering results.

The experimental results demonstrate that using GTM semantic similarity improves the clustering results compared to the baselines. Using Wikipedia concepts as external knowledge also helps increase the accuracy of clustering results. In addition, the Similarity Score Reweighting with Relevance User Feedback strategy achieves the best result and the Internal Centrality-Based Weighting strategy performs better than AVG and MAX, as the first strategy allows users to assign different similarity scores for different elements between two Reddit submissions and the second strategy computes the weighting factor of each element of a Reddit submission based on its semantic content. The Offline–Online clustering algorithm achieves a comparable result to the ICW when the number of concepts is large in the cluster summary.

In future work, we aim to extend the proposed framework to other social network websites, such as Twitter, Facebook, and Google Plus. Another important direction is to extend this framework for studying the spread of rumours in online social networks, for example, visualizing how a rumour-related meme is discussed and spread in Reddit. This will help researchers to understand the patterns of how a rumour is spread, its pattern and detect emerging rumours. Comparing the spread of rumour-driven memes between Reddit and other OSNs and finding a correlation between them will provide a more holistic view of rumour spread.

Acknowledgements The research was funded in part by the Natural Sciences and Engineering Research Council of Canada, International Development Research Centre, Ottawa, Canada, Social Sciences and Humanities Research Council of Canada, CNPq, and FAPESP (Brazil).

References

1. Aggarwal CC, Subbian K. Event detection in social streams. In: SDM conference, vol. 12. Philadelphia, PA: SIAM; 2012, p. 624–35.
2. Aggarwal CC, Han J, Wang J, Yu PS. A framework for clustering evolving data streams. In: Proceedings of the 29th international conference on very large databases, vol. 29. VLDB Endowment; 2003, p. 81–92.
3. Aggarwal CC, Zhao Y, Philip SY. On clustering graph streams. In: SDM conference. Philadelphia, PA: SIAM; 2010, p. 478–89.
4. Banerjee S, Ramanathan K, Gupta A. Clustering short texts using wikipedia. In: Proceedings of the 30th annual international ACM SIGIR conference on research and development in information retrieval. New York: ACM; 2007, p. 787–88.
5. Becker H, Naaman M, Gravano L. Learning similarity metrics for event identification in social media. In: Proceedings of the third ACM international conference on web search and data mining. New York: ACM; 2010, p. 291–300.
6. Berkhin P. A survey of clustering data mining techniques. In: Grouping multidimensional data. Berlin: Springer; 2006, p. 25–71.
7. Bilenko M, Basu S, Mooney RJ. Integrating constraints and metric learning in semi-supervised clustering. In: Proceedings of the twenty-first international conference on machine learning. New York: ACM; 2004, p. 11.
8. Bollegala D, Matsuo Y, Ishizuka M. Measuring semantic similarity between words using web search engines. WWW 2007;7:757–66.
9. Brants T, Franz A. The google web 1t 5-gram corpus version 1.1. Technical Report, 2006.
10. Cataldi M, Di Caro L, Schifanella C. Emerging topic detection on twitter based on temporal and social terms evaluation. In: Proceedings of the tenth international workshop on multimedia data mining. New York: ACM; 2010, p. 4.
11. Caulkins BD, Lee J, Wang M. A dynamic data mining technique for intrusion detection systems. In: Proceedings of the 43rd annual southeast regional conference, vol. 2. New York: ACM; 2005, p. 148–53.
12. Ceccarelli D, Lucchese C, Orlando S, Perego R, Trani S. Dexter 2.0-an open source tool for semantically enriching data. In: International semantic web conference (Posters & Demos). 2014, p. 417–20.

13. Chang JH, Lee WS. Finding recent frequent itemsets adaptively over online data streams. In: Proceedings of the ninth ACM SIGKDD international conference on knowledge discovery and data mining. New York: ACM; 2003, p. 487–92.
14. Chi Y, Wang H, Yu PS, Muntz RR. Moment: maintaining closed frequent itemsets over a stream sliding window. In: Fourth IEEE international conference on data mining (ICDM). Piscataway, NJ: IEEE; 2004, p. 59–66.
15. Chierichetti F, Kumar R, Pandey S, Vassilvitskii S. Finding the Jaccard median. In: Proceedings of the twenty-first annual ACM-SIAM symposium on discrete algorithms. Society for Industrial and Applied Mathematics; 2010, p. 293–311.
16. Choo J, Lee H, Liu Z, Stasko J, Park H. An interactive visual testbed system for dimension reduction and clustering of large-scale high-dimensional data. In: IS&T/SPIE electronic imaging. International Society for Optics and Photonics; 2013, p. 865,402–865,402.
17. Dang A, Makki R, Moh'd A, Islam A, Keselj V, Milios EE. Real time filtering of tweets using wikipedia concepts and google tri-gram semantic relatedness. In: Proceedings of the TREC, 2015.
18. Dang A, Moh'd A, Gruzd A, Milios E, Minghim R. A visual framework for clustering memes in social media. In: Proceedings of the 2015 IEEE/ACM international conference on advances in social networks analysis and mining. New York: ACM; 2015, p. 713–20.
19. Dang A, Michael S, Moh'd A, Minghim R, Milios E. A visual framework for clustering memes in social media. In: Proceedings of the 2016 IEEE/ACM international conference on advances in social networks analysis and mining. New York: ACM; 2016.
20. Dang A, Moh'd A, Milios E, Minghim R. What is in a rumour: combined visual analysis of rumour flow and user activity. In: Proceedings of the 33rd computer graphics international. New York: ACM; 2016, p. 17–20.
21. Deza MM, Deza E. Encyclopedia of distances. Berlin: Springer; 2009.
22. FEMA: Hurricane sandy: Rumor control @ONLINE, 2012. Available: http://neteffect. foreignpolicy.com/posts/2009/04/25/swine_flu_twitters_power_to_misinform. Accessed 15 April 2015.
23. Giannella C, Han J, Pei J, Yan X, Yu PS. Mining frequent patterns in data streams at multiple time granularities. Next Generation Data Mining. 2003;212:191–212.
24. Gouda K, Zaki M. Efficiently mining maximal frequent itemsets. In: Proceedings IEEE international conference on data mining (ICDM). Piscataway, NJ: IEEE; 2001, p. 163–70.
25. Halkidi M. Quality assessment and uncertainty handling in data mining process. In: EDBT Ph.D Workshop, 2000.
26. Heylighen F. What makes a meme successful? Selection criteria for cultural evolution. In Proceedings 15th International Congress on Cybernetics, 1998.
27. Hong L, Davison BD. Empirical study of topic modeling in twitter. In: Proceedings of the first workshop on social media analytics. New York: ACM; 2010, p. 80–8.
28. Hu X, Zhang X, Lu C, Park EK, Zhou X. Exploiting wikipedia as external knowledge for document clustering. In: Proceedings of the 15th ACM SIGKDD international conference on knowledge discovery and data mining (KDD). New York: ACM; 2009, p. 389–96.
29. Hu M, Liu S, Wei F, Wu Y, Stasko J, Ma KL. Breaking news on twitter. In: Proceedings of the SIGCHI conference on human factors in computing systems. New York: ACM; 2012, p. 2751–54.
30. Islam A, Milios E, Kešelj V. Text similarity using google tri-grams. In: Proceedings of the 25th Canadian conference on advances in artificial intelligence, Canadian AI' 12. Berlin/Heidelberg: Springer; 2012, p. 312–17.
31. JafariAsbagh M, Ferrara E, Varol O, Menczer F, Flammini A. Clustering memes in social media streams. Soc Netw Anal Min. 2014;4(1):237.
32. Kwak H, Lee C, Park H, Moon S. What is twitter, a social network or a news media? In: Proceedings of the 19th international conference on world wide web. New York: ACM; 2010, p. 591–600.
33. Lee H, Kihm J, Choo J, Stasko J, Park H. iVisClustering: an interactive visual document clustering via topic modeling. In: Computer graphics forum, vol. 31. Wiley Online Library; 2012, p. 1155–164.

34. Leskovec J, Backstrom L, Kleinberg J. Meme-tracking and the dynamics of the news cycle. In: Proceedings of the 15th ACM SIGKDD international conference on knowledge discovery and data mining. New York: ACM; 2009, p. 497–506.
35. Li HF, Lee SY, Shan MK. An efficient algorithm for mining frequent itemsets over the entire history of data streams. In: Proceedings of first international workshop on knowledge discovery in data streams, 2004.
36. Mannila H, Toivonen H, Verkamo AI. Discovery of frequent episodes in event sequences. Data Min Knowl Disc. 1997;1(3):259–89.
37. Morozov E. Swine flu: Twitter's power to misinform @ONLINE, 2009. Available: http:// neteffect.foreignpolicy.com/posts/2009/04/25/swine_flu_twitters_power_to_misinform. Accessed 15 April 2015.
38. Pedersen T, Patwardhan S, Michelizzi J. Wordnet::similarity: measuring the relatedness of concepts. In: Demonstration papers at HLT-NAACL, HLT-NAACL–demonstrations. Stroudsburg: Association for Computational Linguistics; 2004, p. 38–41.
39. Pramod S, Vyas O. Data stream mining: a review on windowing approach. Global J Comput Sci Technol Softw Data Eng. 2012;12(11):26–30.
40. Strehl A, Strehl E, Ghosh J, Mooney R. Impact of similarity measures on web-page clustering. In: Workshop on artificial intelligence for web search. AAAI; 2000, p. 58–64.
41. Thom D, Bosch H, Koch S, Wörner M, Ertl T. Spatiotemporal anomaly detection through visual analysis of geolocated twitter messages. In: Pacific visualization symposium (PacificVis). Piscataway, NJ: IEEE; 2012, p. 41–8.
42. Zhao Y, Yu P. On graph stream clustering with side information. In: Proceedings of the seventh SIAM international conference on data mining. Philadelphia, PA: SIAM; 2013, p. 139–50.

A System for Email Recipient Prediction

Zvi Sofershtein and Sara Cohen

1 Introduction

Email is one of the most popular means of communication, and its volume seems to be consistently increasing. It is estimated that in 2015, approximately 205 billion email messages will be sent per day [1]. However, email use is fraught with potential serious pitfalls, with one big problem being mistakenly sending an email to the incorrect recipient. Online searches about email blunders yield a large number of such unfortunate occurrences. In fact, one real-life blunder even ended up being the inspiration for a 2011 Superbowl commercial dealing with precisely this mistaken recipient problem.

As email is such a common means of communicating, and email mistakes are potentially disastrous, systems that can help reduce email error are of importance. This article considers the problem of accurately predicting recipients for a given email. Once such recipients can be predicted, the system can alert the user if he seems to be sending the email to the wrong persons. In addition, email recipient prediction is useful as a GUI feature, as it allows for easy and automatic selection of recipients, reducing the need for manual typing of this information.

User behaviour in emails has been studied in many previous works. Email content was used in [2] for personality prediction. In [3, 4], evaluations were performed to determine the factors that cause users to take actions on their incoming mails. These actions included replying, saving and marking as important. A similar problem was studied in [5], as well as in [6], which also attempted to predict response time to incoming mails. In [7], a model of email behaviour was proposed to characterize

Z. Sofershtein • S. Cohen (✉)
The Rachel and Selim Benin School of Computer Science and Engineering,
Hebrew University, Jerusalem, Israel
e-mail: zvisofer@gmail.com; sara@cs.huji.ac.il

© Springer International Publishing AG 2017
M. Kaya et al. (eds.), *From Social Data Mining and Analysis to Prediction and Community Detection*, Lecture Notes in Social Networks,
DOI 10.1007/978-3-319-51367-6_2

email users. Such individual-user based analysis can help detect abnormal email activity, and form a defense against the spread of worms and spam.

In the past, other systems were developed to protect from human errors in emails. For example, [8, 9] proposed systems that detect missing attachments in emails. Such a system was added to *Gmail*, currently one of the most popular email services in the world. We believe that other, intelligent automatic assistants will be added in the future to email services. A recipient prediction system can be a valuable automatic assistant for email services. It can also help detecting much more dangerous errors than missing attachments, e.g. email leaks.

The problem of email recipient prediction has been studied in the past and experimentation has focused on the Enron dataset [10]. It was pointed out in [11] that at least 20.5% of Enron users were not included as recipients in some mails, even though they were intended recipients, i.e., mistakes in determining email recipients are commonplace. In [11–14] systems were presented that combine textual features with recency and frequency features. However, the textual features used in these works were rather limited, using off-the-shelf text similarity metrics, and experimentation was confined to the Enron dataset. In [15, 16], a recipient prediction method, based on social network features, was proposed. Intuitively, their method was based on a recipient seed in the composed message and past co-occurrence of possible additional recipients with the seed. However, this approach does not reduce the effort of inserting the seed, nor is it applicable when there is only one recipient.

In this article we present a system that can accurately predict email recipients for a new email, given the user's email history. The system computes a variety of features, that are either temporal or textual based, and uses these features to learn a ranking function over the user's contacts, given a new email. Our textual features include a more targeted approach (based on email greetings) to improve the effectiveness of the system. Extensive experimentation over personal, business and political emails has shown the effectiveness of the system, which currently works for both English and Hebrew language emails.

Our system was discussed and demonstrated in [17]. In the demonstration, we exhibited two applications of our system. First, we demonstrated recipient auto-complete. Second, we built an alerting mechanism, which is triggered when the user is attempting to send an email to unlikely recipients. These are only a few of the possible applications of this important capability.

This work is a significant extension of [17], which was a short version, containing few details. In particular, this article contains extensive discussion on the following topics (which were either completely lacking in [17], or were only briefly touched upon):

- the greeting feature, including its derivation and experimental evaluation of its quality (performed on two languages—English and Hebrew—and over three domains);
- the implementation of the feature extractor, including a complexity analysis for both learning and ranking;

- comprehensive experimental evaluation, including three new test settings:

  - cross-user testing, in which ranking functions that were trained on some users are tested on other users. (This configuration is very useful when there is limited data on a user, and training a ranker on that data is not possible.) The cross-user setting is evaluated for both within-domain and cross-domain configurations.
  - evaluation when there are known recipient prefixes
  - analysis of feature-performance

- a comparison with results shown in [14], showing the superiority of our methods.

Section 2 begins by defining the problem of interest. We depict the features used in our system in Sect. 3. Section 4 focuses on the greeting extraction method and includes an experimental evaluation. Section 5 explains the learning framework in which the discussed features are used and presents our implementation of the feature extractor component. Section 6 presents the extensive experimental evaluation performed in this work.

2 The Recipient Prediction Problem

In this section, we formally define the recipient prediction problem. We also refer to the accompanying ranking evaluation problem.

The problem at hand can be defined as follows. We consider a user u who has, until some point in time t, sent and received emails. Let $C_{u,t}$ be the *contact list* of user u at time t, i.e., the set of people who have either sent an email to u, or been sent an email by u, before time t. Now, given a new email message m that u composes at time t, we would like to predict the recipient(s) to whom u intends to send this specific email. More precisely, our goal is to define a ranking function $r_m : C_{u,t} \rightarrow \mathbb{R}$, such that for contacts $c_i, c_j \in C_{u,t}$

$$r_m(c_i) > r_m(c_j)$$

implies that it is more likely that m will be sent to contact c_i, than to contact c_j. We say that an email is *predictable* if at least one of its true intended recipients is in $C_{u,t}$. For obvious reasons, we are only interested in predictable emails (and all experimentation presented in Sect. 6 is for such emails). Figure 1 shows an example of predictable and non-predictable mails.

To summarize, a recipient prediction system S takes as input a mailbox $\mathcal{M}$, and a message m where the recipients have not been assigned yet. Then, S outputs a ranking function $r_m : C_{u,t} \rightarrow \mathbb{R}$. That is, $S_{\mathcal{M}}(m) = r_m$.

Note that evaluating the quality of a ranking is a non-trivial task. Given a message m, and the outputted ranking $r_m = S_{\mathcal{M}}(m)$, we wish to evaluate the quality of r_m. Naturally, r_m is optimal if and only if for any recipient $c_i \in C_{u,t}$ of the message m,

From:	John
To:	Jane, Mark
Subject:	On the procurement of money

(a) **Non-predictable w.r.t. John**. The first email in a mailbox can never be predictable, as the contact list is empty.

From:	Martha
To:	John
Subject:	Trip to Portugal

(b) **Non-predictable w.r.t. John**. We only predict outgoing emails, while this one is an incoming email.

From:	John
To:	Chris, Jane
Subject:	Monetization strategy for free apps

(c) **Predictable w.r.t. John**. Jane has been sent an email by John (it is sufficient to have one existing contact among the recipients).

From:	John
To:	Martha
Subject:	Flight tickets

(d) **Predictable w.r.t. John**. Martha has sent an email to John. Note there are no outgoing mails to Martha.

From:	John
To:	Robert
Subject:	Exam appeal

(e) **Non-predictable w.r.t. John**. Robert has neither sent an email to John, nor he has been sent an email by John.

Fig. 1 The mailbox of John consists of five emails, which are ordered here chronologically. The predictability of each email is explained in the respective sub-caption ((**a**)–(**e**))

and for any non-recipient $c_j \in \mathcal{C}_{u,t}$, it holds that $r_m(c_i) > r_m(c_j)$. In other words, an optimal ranking has all recipients on top, followed by all non-recipients.

However, given two rankings r_m and r'_m which are not optimal, there might be no natural way to determine which of them is better. For example, consider the scenario shown in Table 1. In this example, neither of the rankings is strictly better than the other. Moreover, the relative quality highly depends on the application of

Table 1 Two possible
rankings for a mail sent to
Jane and Mark

Contact	Is a recipient?	r_m	r'_m
Jane	Recipient	4	3
Mark	Recipient	1	2
Chris	Non-recipient	3	4
Martha	Non-recipient	2	1

the system. If the application auto-completes according to the top ranked contact, then r_m is better. On the other hand, if the applications alerts when a message is sent, presumably by mistake, to the lowest ranked contact, then r'_m is better.

For emails that have a single recipient, it seems easy to determine which ranking is better. The ranking r_m is strictly better than the ranking r'_m, if and only if the correct recipient is ranked higher in r_m than in r'_m. Yet, even in a single recipient case, similar difficulties arise when trying to compare ranking systems on multiple emails.

In order to evaluate our ranking functions, we use measures that are commonly applied for general ranking problems (MRR and top-k). A side note here is that these measures are not a perfect fit for our problem. First of all, MRR was designed for search engines, which have enormous amount of elements to rank. In the recipient prediction problem, the amount of contacts that are ranked is typically much smaller. In addition, both MRR and top-k are focused on the top of the ranking. In some possible applications of the recipient prediction problem, though, the bottom of the ranking is of importance, e.g., when alerting the user when she has typed unlikely recipients. Nevertheless, we refrain from defining new statistical measures, and prefer to use the standardized measures.

We formally define the measures MRR (mean reciprocal rank) and top-k. Let q_i be a query, and let r^i be a ranking for q_i. We denote by c_i the set of correct answers for q_i. The reciprocal rank of r^i, denoted by RR_i, is the multiplicative inverse of the rank of the first correct answer. The *mean reciprocal rank* is the average of the reciprocal ranks of results for a sample of queries Q.

$$\mathrm{MRR}(Q) = \frac{1}{|Q|} \sum_{i=1}^{|Q|} \mathrm{RR}_i = \frac{1}{|Q|} \sum_{i=1}^{|Q|} \frac{1}{\min_{1 \le j \le |r_i| : r^i_j \in c_i} j}$$

The top-k measure is defined to be the percentage of rankings for a sample Q, where there is at least one correct answer in the top-k ranked elements. Obviously, this measure is monotonically increasing with k.

3   Features for Recipient Prediction

In this section we define all features used by our recipient prediction system. In the following, we fix a user u. The function r_m will be learned, for users $c \in \mathcal{C}_{u,t}$, from the history of user u by leveraging a variety of features that are either temporal or textual. All features are normalized to a range of $[-1, 1]$ to ensure proper convergence of the learning mechanism applied and to adjust the values so that functions learned from different users will be comparable.

3.1   Temporal Features

We consider four features that are based on the time in which message m is sent. In the following, we use $\mathcal{M}_{c_1 \rightarrow c_2, (t_1, t_2)}$ to denote the set of messages sent by c_1 to c_2 in the time interval between time t_1 and t_2 (excluding). Note that we will use the wildcard $*$ instead of c_1 or instead of c_2 to denote, respectively, the set of *all* emails sent to c_2 or sent by c_1 within the given time interval. Also note that we use 0 to denote the beginning of time, and thus, an interval of the form $(0, t_2)$ will allow for all messages until time t_2.

The first two features return values, for a contact c, that are proportionate to the frequency of messages written by u to c, or by c to u.

- **Outgoing Message Percentage:** The *outgoing message percentage* of c for u at time t is the percentage of messages sent by u to c, with respect to all messages written by u, i.e.,

$$\frac{|\mathcal{M}_{u \rightarrow c, (0,t)}|}{|\mathcal{M}_{u \rightarrow *, (0,t)}|}$$

- **Incoming Message Percentage:** Similarly to the previous feature, the *incoming message percentage* is the percentage of messages sent to u by c, with respect to all messages sent to u, i.e.,

$$\frac{|\mathcal{M}_{c \rightarrow u, (0,t)}|}{|\mathcal{M}_{* \rightarrow u, (0,t)}|}$$

The other two features are based on the time elapsed since u sent an email to c or received an email from c. Instead of using this time value directly as a feature, we choose to count how many messages have been sent (or received) since the last communication with c. This is an important normalization method as different users can have different habits—some send out emails every few seconds, while for others, hours may pass between consecutive emails.

- **More Recent Outgoing Percentage:** Let t' be the time at which the last message was sent from u to c. The *more recent outgoing percentage* of c at time t is the percentage of messages sent by u since she last emailed c, i.e.,

$$\frac{|\mathcal{M}_{u \rightarrow *, (t',t)}|}{\alpha |\mathcal{M}_{u \rightarrow *, (0,t)}|}$$

If u has never written an email to c, this feature has the constant value 1. Note the use of α in the denominator to further differentiate (and increase the gap) between the cases in which a message was sent to c only very early on, and the case in which no message was ever sent to c.

- **More Recent Incoming Percentage:** Let t' be the time at which the last message was sent from c to u. The *more recent incoming percentage* of c at time t is the percentage of messages sent to u since she last received an email from c, i.e.,

$$\frac{|\mathcal{M}_{* \rightarrow u, (t',t)}|}{\alpha |\mathcal{M}_{* \rightarrow u, (0,t)}|}$$

As before, if u has never received an email from c, this feature has the constant value 1.

3.2   Textual Features

We consider three features that are based on the textual content of the message m being sent. Note that we include both the subject and the content of the message, when considering the text of m. The first two features are used to indicate whether contact c has either received from u, or sent to u, a message with text similar to that of m.

- **Outgoing Textual Similarity:** We index all previous emails sent to u, or received by u, using the Apache Lucene search engine.[1] Then, given a new message m, we search the indexed data to find similar messages sent by u. Let $\mathcal{M}^m_{u \rightarrow *, (0,t)}$ be messages similar to m, sent by u, found using the Okapi BM25 textual similarity measure with Lucene. Then, for a given contact c, the *outgoing textual similarity* feature is 1 if $\mathcal{M}^m_{u \rightarrow *, (0,t)}$ includes a message sent to c, and -1, otherwise.
- **Incoming Textual Similarity:** This feature is similar to the previous. Let $\mathcal{M}^m_{* \rightarrow u, (0,t)}$ be the most similar messages to m, sent to u, according to the Okapi BM25 textual similarity measure. Then, for a given contact c, the *incoming textual similarity* feature is 1 if $\mathcal{M}^m_{* \rightarrow u, (0,t)}$ includes a message sent by c, and -1, otherwise.

---

[1] http://lucene.apache.org/

We note the underlying assumption here is that the interaction between two users may be on diverse topics, and it is sufficient to find a single match for a contact c, to derive a strong indication that c should be the message recipient. Previous works measured the average similarity of the messages between the user and the candidate, but this feature degrades for contacts that are very popular, and hence, appear in many contexts.

Greeting Our final textual feature takes into consideration the fact that emails often include, in their content, the name, or nickname, of the recipient, by way of a *greeting* at the beginning. For example, an email may begin with text such as Dear Sam and Jim, Hi Mom, or Sweetheart! In each of these cases, the user has included some indication as to the recipient (i.e., Sam, Jim, Mom and Sweetheart). To the best of our knowledge, this feature has not been considered in the past for email recipient prediction. Due to its complexity, this feature will be discussed at length in the next section.

4 Finding Names in Greeting

As mentioned in Sect. 3, the greeting feature is designed to capture a structure of a greeting in a mail's content, and to extract names or nicknames of recipients from it.

To leverage email greetings, we associate with a message m the set $\mathcal{G}_m$ of all names or nicknames appearing in the greeting of m. We note that this set can be empty, if there is no greeting, or if the greeting does not contain any names (such as Hi All!). We explain below how to automatically create the set $\mathcal{G}_m$ using a rule-based system.

We define the value of the *greeting* feature given a new message m and a contact $c \in \mathcal{C}_{u,t}$ as:

$$\begin{cases} 0 & \mathcal{G}_m = \emptyset \\ 1 & \mathcal{G}m \cap (\bigcup_{m' \in \mathcal{M}_{u \to c,(0,t)}} \mathcal{G}_{m'}) \neq \emptyset \\ -1 & \text{otherwise} \end{cases}$$

In other words, if m contains no greeting names, then the feature has the constant value of 0 for all contacts. Otherwise, m has some greeting names. In this case, the feature has the value of 1 for contacts who have been sent emails from u using some greeting from $\mathcal{G}_m$, and -1 for the remaining contacts.[2]

---

[2]Some emails have several recipients and more than one name in the greeting. In such cases, we cannot distinguish to which recipient each name refers, and therefore $\mathcal{G}_m$, for an email sent to c, may actually contain a name not referring to c.

4.1 Extracting Names from a Greeting

We now discuss our rule-based system for deriving $\mathcal{G}_m$ from m. This system is not foolproof, but has been experimentally evaluated over large corpora (both in English and in Hebrew) and has been shown to be very effective. Here we present the features of our system.

The basic assumption behind our rules is that greetings tend to be short and mostly comply with one of several popular structures. Typically, a greeting will appear at the beginning of the message, before any separating symbol (such as a period, comma or colon). An exception to this rule is for messages with multiple recipients, in which case separator symbols may appear between names (e.g., Bob, Jim, Sally:). We do not discuss this case any further, but special rules have been developed that are applicable for messages with multiple recipients.

We differentiate between five types of tokens (i.e., textual sequences) that can appear at the beginning of a message, until the first separator:

1. *Greeting words* such as hi, dear, mr., thanks, etc.
2. *Nonspecific names* such as there, again, all, etc. All of these are commonly used in the same place where a regular name or nickname could be used.
3. *Standalone words* such as yes, sure, oops, etc. (These are needed, e.g., to differentiate between a message beginning Jim! and one beginning Oops!.)
4. *Greeting names* such as David, Honey, etc. These are words not belonging to the previous groups.
5. *Connectors* such as or, &, etc.

Some examples of extraction rules using these groups, as well as text that is correctly parsed, appear in Fig. 2. We note that our system correctly recognizes that the following phrases contain no names or nicknames: Hi, thanks!, hi there, hello again, Yes., sure....

Now that we covered the basic mechanism we will discuss few more enhancements and special cases. First, we consider an additional type of tokens, namely *Titles*. Titles include honorifics, such as Miss, Sir, Ms, Mrs, Dr, etc. They also

Rule	Text
\<Greeting name\>	David, honey
\<Greeting word\>\<Greeting name\>	Dear David. Hi dear!
\<Greeting name\>\<Connector \>\<Greeting name\>	David or Mike:
\<Greeting word\>\<Greeting name\> \<Connector \>\<Greeting name\>	Hello David & Mike

Fig. 2 Several of our parsing rules, and examples of text that is successfully parsed using these rules

include other titles and positions such as `Professor, General, Minister,` etc. These titles can be typically used in two different forms. The first form is before the actual name of the title subject, such as `Mrs Robinson -, Professor Layton:` etc. The second form is where the title appears alone without any particular name (proper noun), such as `Dear Sir,, Hello Professor,` etc.

The way we should handle titles is not trivial. Basically, we should decide if we wish to extract titles, or to ignore them. We use a hybrid approach, where in some cases we extract the title, while in other cases we ignore it. When the greeting uses the first form, i.e. a title followed by a specific name, we ignore the title and extract only the name itself. For example, given `Mrs Robinson:`, we extract `Robinson`. When the greeting uses the second form, i.e. only a title, we extract the title. For example, given `Dear Sir,`, we extract `Sir`.

In order to justify our approach, we will claim that the name itself is usually more indicative than the title. This is because a title is usually less specific and is applicable towards a larger group of persons. Name collisions, on the other hand, are rarer than title collisions. For this reason, when we have a specific name we can ignore the title and still maintain most of the information within the specific name. On the other hand, in the second form where we only have a title, it is meaningful to extract it. First of all, the title discloses something about the title subject. It can be the gender, family status, education, position, etc. This is due to the applicability of the title. For example, `Mrs` is not applicable towards any person who is not a married female. Another example is that `Dr` is not applicable towards any person who does not hold a PhD. Secondly, the title preference adds even more information. The fact that a certain title is applicable towards a certain person, does not mean it will be used. For example, the title `Sir` is unlikely to be used to greet a family member or a close friend, even if they are a male. We can see that extracting a title can help us identify the recipient, or at least narrow down the candidates.

Yet, this approach is not foolproof, and has several drawbacks. One of which is name collisions. In this case it is not clear that we need to ignore the title. For example, `Mrs Robinson` and `Mr Robinson` are very unlikely to be used towards the same person. On the other hand, `Mrs Robinson` and `Dr Robinson` might be used towards the same person, or might describe two different persons. As we wish to avoid defining a complete taxonomy of titles, we prefer to simply ignore the title in such cases. Note that while a complete taxonomy can help differentiating between `Mrs Robinson` and `Mr Robinson`, it is useless in differentiating between `Mrs Robinson` and `Dr Robinson`, where more prior knowledge is required and occasional errors are inevitable. Fortunately, name collisions are fairly rare. Our approach also ensures that we detect the correct recipient, as well as a subset of contacts with the same name. As this subset is much smaller than the set of all contacts, the contribution of the greeting features remains substantial, even in such difficult cases.

Another drawback is when both forms of titles are used on and off. For example, if we send a first mail to a contact starting with `Professor Layton:`, and later send him a second mail starting with `Dear professor,`. As we have already defined, from the first mail we will extract `Layton`, while from the second mail we will extract `professor`. As a result, our greeting feature will not detect any connection between these two mails. Yet, it is worthwhile mentioning that `Dear professor,` can be also used to greet other professors in our contact list, not only professor Layton.

The reader may have noticed that we did not include *Titles* in the extraction rules. Adding *Titles* to the extraction rules explicitly will force us to write twice as many rules. That is, for every rule, there will be a complementary rule where a *Title* appears before the extracted name. A simple alternative is to use the following wrapping rule. Given a text T, create a text T' where any *Title* token has been removed. Then, try to extract greeting names from T'. If any have been found, return them. If no names were extracted from T', try to extract names from T, and return any names found.

Our greeting extractor also needs to determine whether or not the text it is processing is a natural language text. For instance, many mails consist of a single internet link. In such cases, the sole purpose of the mail is to share the content of the link, and of course, no greeting is being used. We wished to avoid applying our extraction rules on internet links, since they might extract invalid names. For example, applying the extraction rules on `dropbox.com` will output the token `dropbox`. In order to avoid such errors, we test if the mail begins with a link. If it does, we do not return any name. A similar solution can be applied towards other types of non-natural language text, such as *HTML* tags. However, in our development set, all mails started either with natural language text, or with internet links.

There are few problematic issues that our greeting extractor does not correctly handle. From our empirical study, we have seen that these issues are rare. Because of the complexity of handling these issues, trying to correctly deal with these cases may even increase the error rate.

The first issue is typographical errors. When these errors take place, our greeting extractor will simply extract the misspelled names, and the feature will not indicate any connection with the respective, correctly spelled names. An alternative approach could be measuring edit distance between names instead of testing them for absolute identity. Such an approach can also use a weighted edit distance, based on letter resemblance and distance on a standard keyboard. On the other hand, it could cause errors for similar names, e.g., `Jake`, `Jane`, `Jade`. The enormous number and variety of given names is a major drawback for this alternative approach.

Another issue our greeting extractor does not handle is multi-token names. When a multi-token name is used, i.e. `Bill Gates:`, the extractor will not extract any name. This takes its toll on the extractor's recall, but is necessary to keep the precision high. Note that while the number of popular `Standalone words` is limited, the number of word couples that can form a sentence is enormous. If the extractor were to extract two tokens, it would also extract incorrect names, such as `Go home` from `Go home,`. A naive approach that tests whether these tokens are English words will also often fail. For example, `Bill` and `Gates` are perfectly valid English words.

4.2 Handling Additional Languages

In this part we discuss the possibility of adding more languages to the greeting name extractor. As seen, unlike text similarity, the greeting name extraction is language-dependent. Handling more languages requires finding words for each token set. In addition, it requires finding more greeting patterns, that may be specific for the given language. Once the popular greeting patterns are found, additional parsing rules for these patterns should be composed.

To demonstrate the addition of a language to the greeting extractor, we will briefly discuss the support of the Hebrew language, that we added to our extractor. For Hebrew, the basic framework remains the same. We used the same extraction rules that were previously presented, and added a few additional rules. The major modification is that the token sets used in the rules, i.e. *Greeting words*, *Nonspecific names*, *Standalone words*, *Greeting names*, *Connectors* and *Titles*, were expanded to also contain the Hebrew tokens that belong to each set. The advantage of this approach is that the extractor does not need to detect which language it is parsing. Another advantage is that this framework should work fairly well, even if we do not add new rules. This is because many of the greeting patterns are the same for both languages, and can be extracted using the same rules. Yet, good adaptivity of the framework from English to Hebrew means that the framework is likely to adapt well to other languages, especially languages which are more similar to English than Hebrew. While the resulting extractor achieved similar precision in Hebrew, it had significantly lower recall. The reason was that there are additional popular greeting patterns in Hebrew. For example, a *Greeting name* followed by a *Greeting word*. Therefore, additional parsing rules were added to capture this pattern and other, similar patterns.

The effectiveness of the extractor may vary from one language to other. Even if the same effort has been made for two languages, the performance of the extractor can be very different on these languages. One language might have more complex and diversified greeting patterns, or might be more complicated to parse. In the next subsection we will, among other things, compare between the performance on Hebrew texts and on English texts. Such a comparison can teach us about the adaptivity and challenges of name extraction on different languages, using our extraction method. Generally, we cannot argue that our extraction framework will work well in any language.

4.3 Experimental Evaluation

In this subsection, we experimentally evaluate the greeting feature. More precisely, this evaluation does not measure the accuracy of the greeting feature itself, but rather the accuracy of the method which extracts greeting names from a mail. For this purpose, we used a diverse dataset of mails, containing mails from the personal (Gmail), business (Enron) and political sectors. More details on this dataset appears in Sect. 6.

Fig. 3 Number of tagged
mails from each dataset

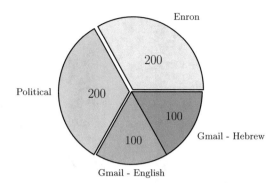

Gmail - English

Table 2 Percentage of mails
containing greeting with
names, per dataset

Dataset	Named greeting (%)
Enron	50
Political	45
Gmail-english	49
Gmail-hebrew	48

For this evaluation, 600 random mails from our datasets were tagged manually. For each mail m, the set $\mathcal{G}_m$ was tagged, i.e. all names appearing in the greeting of the mail. As defined in Sect. 3, this set is empty when there is no greeting or if the greeting does not contain any specific names.

To form a diverse greeting dataset, an equal number of mails were taken from each mail dataset. For the Gmail dataset which contained both English and Hebrew mails, an equal number of mails were taken for each language. Figure 3 summarizes the distribution of tagged data.

The tagged mails were sampled from both incoming and outgoing mails of the mailboxes. Note that the feature extraction described in previous subsections was performed only on outgoing mails. Including incoming mails in this evaluation allowed us to obtain even more diversified mail set. This was specifically substantial in the Gmail dataset. In this dataset, due to privacy concerns, all tagged mails were taken from a single mailbox. In this case, using only outgoing mails would have resulted in mails composed by a single user. Therefore it was crucial to include incoming mails as well.

Another important statistic obtained from the tagged dataset is the percentage of mails that contain a greeting with names. The data for labelling was sampled randomly, and the manual tagging indicates whether the mail contains a greeting with names or not. Thus, for each domain, this is an estimate of the overall percentage of mails containing a greeting with names. As shown in Table 2, roughly half of the mails from every dataset contain a greeting with names. This is encouraging, since it shows that even when considering an informal mail collection such as the Gmail dataset, greetings with names remain commonplace.

Figure 4 shows the accuracy of our method for every data set. The measures shown are recall, precision and F1 score (the harmonic mean of recall and precision).

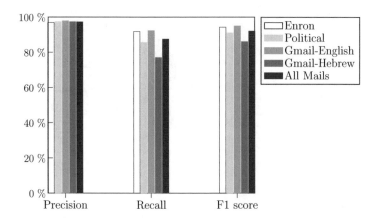

Fig. 4 Accuracy of the proposed greeting names extraction method on different domains

Table 3 A demonstration of recall and precision calculation across mails

	m_1	m_2	m_3	m_4
Correct names	{David, Mike, Josh}	{Sally}	{John}	{}
Extracted names	{David, Harry}	{Sally}	{}	{Don}

	m_1 (%)	m_2 (%)	m_3	m_4	Total (%)
Precision	50	100	N/A	0%	50
Recall	33	100	0%	N/A	44

Observe that the precision is consistently higher than the recall. The overall F1 score is 92.1%, embodying a precision of 97.3% and recall of 87.5%.

The precision and recall are calculated separately for each mail, then averaged across all mails. To demonstrate this, consider the example calculation given in Table 3. Obviously, mails that do not contain greeting names are excluded from the recall calculation. Similarly, mails for which no names were extracted are excluded from the precision calculation. An alternative approach where the recall and precision are averaged across all greeting names will erroneously give more weight to mails which contain more greeting names.

Comparison between languages shows that the performance on Hebrew, in particular the recall, is considerably lower than on English. This is not surprising since the Hebrew language is considered harder to process than English. Even tokenization becomes a complicated task in Hebrew, since some prepositions and conjunctions are concatenated with the successive word. In addition, word order is much less strict in Hebrew, which increases the number of greeting patterns.

In the English datasets, the lowest recall was observed in the Political dataset. The false-negatives contain mostly multi-token names, such as Dear Governor Bush:. Such simple cases can be handled by adding more titles to the title collection. Yet, any new domain can introduce new titles and these titles may have multiple forms and abbreviations (i.e. Dear Gov. Bush,). Another set of multi-token names is simply full names (i.e. `Dear Governor Jeb Bush,`) where our method will not extract the name to avoid the risk of false-positives.

Going over the false-positives shows mostly new standalone words that were not taken into account in our method (i.e. `SALARIES, SALARIES, SALARIES!!!`). Yet these words are extremely rare as the overall precision is 97.3%.

5   Implementation

Given the features, we learned a ranking function using a ranking SVM [18]. We assume the reader already has some familiarity with the concept of an SVM. Very much like a standard SVM, a ranking SVM learns a linear function of the features. The difference between the two is the representation of the examples, as well as the usage of the learned function. As opposed to a standard SVM where each data unit is a single example, ranking SVMs process sets of examples having a well-defined ranking among themselves. Naturally, these rankings are only known for tagged data, therefore a ranking SVM is used to learn how to rank untagged sets of examples. For this purpose, the linear function learned is used to calculate a score for each example in the set. Finally, the ranking of the set is induced by the scores of the examples.

The basic structure of our learning flow is shown in Fig. 5. In this flow, we extract features from a set of predictable mails. The set of feature vectors for each mail constitutes a query. Our rule for determining the rank of each feature vector is boolean. If the respective mail was sent to the respective contact, the value is 1. Otherwise, the value is -1. Note that this binary ranking gives the desired ranking. A perfect contact ranking has all the recipients on top, followed by all non-recipients. The inner order within each of these two sequences may be arbitrary, and therefore, all the members of each sequence have the same rank value.

An important enhancement to this simple approach was to limit the number of negative examples, i.e. feature vectors produced from non-recipients. Mailboxes usually have dozens or hundreds of contacts, but most mails have only a few recipients. Therefore, taking all possible non-recipient contacts will result in a training set that consists almost entirely of negative examples. Since an SVM attempts to minimize the error on the training set, when applying it to such datasets, it will learn to always classify as negative and will output an arbitrary ranking function. To avoid this problem, we limited the number of negative examples of

Fig. 5 The learning flow that produces a ranking function

Fig. 6 The contact ranking flow

each mail by the number of its positive examples. These examples were randomly selected from all possible negative examples of the email.

Given a new email, for which we would like to rank possible contacts, ranking proceeds as depicted in Fig. 6. We extract features from the single new mail, to produce a feature vector for each contact. Then, we apply a previously trained ranking function, to rank all feature vectors. Finally, we derive the ranking of contacts, which are associated with each feature vector.

As is apparent from Figs. 5 and 6, a critical component in our system is the *feature extractor*. During the learning stage, we use an SVM implementation that runs in linear time in the size of the examples. While ranking, we simply compute the value of a classifier function over the contact feature vectors, which again is efficiently performed. Hence, in the remainder of this section, we focus on making feature extraction efficient, in order to reduce the time for the main bottleneck of the system. Note that this is particularly critical during ranking, as the ranking should be performed faster than the time it takes to insert the address manually. Moreover, for a good user experience it should be perceived as if the recommendations appeared immediately.

5.1   Definitions and Assumptions

Our input is a single mailbox of a given user. We will use the following notations for the input size:

- **Number of contacts:** n_c.
- **Number of outgoing mails:** n_o.
- **Number of incoming mails:** n_i.
- **Average number of recipients per outgoing mail:** a_r.
- **Number of distinct terms that appear in all mail texts:** n_t.

We consider the number of features to be a constant. Throughout this work, the number of features never exceeds 7, and does not depend on the input size.

In addition, we assume that the number of distinct names used to greet each contact is constant. Note that this aligns with the underlying assumption of the greeting feature, which assumes that the names used to greet a certain contact repeatedly appear in many mails. The whole assumption about names and nicknames is that they seldom change, and are used over and over again.

We also assume the length of a single mail's content is constant. In particular, we assume that the length of a single thread is constant. Note that an email typically contains the whole thread in its content.

5.2 Implementation

The implementation uses several indexes that allow each feature to be calculated efficiently.

- **Temporal indexes for outgoing and incoming mails:** We store two search trees, one for incoming mails and one for outgoing mails. The trees are organized based on mail timestamps. Using these structures, all outgoing and incoming mails for a given time period can be found.
- **Contact indexes for outgoing and incoming mails:** We store two hash indexes, one for incoming mails and one for outgoing mails. These hash indexes associate each contact name with a search tree containing all incoming (respectively, outgoing) mails, organized based on mail timestamps. Note that a mail can appear in multiple entries of the contact index for outgoing mails (as it can be sent to multiple recipients), but will appear once in the contact index for incoming mails.
- **Per-contact greeting names:** We store a hash index that maps each contact to the set of all the greeting names used to in outgoing mails sent to the contact.
- **Textual index:** Finally, we used Lucene to store all mail contents. This is used to compute the textual similarity feature.

See Fig. 7 for an illustration of some of the index structures.

Clearly, these structures can be built efficiently. Lucene uses space that is proportional to the total mail contents. The additional five indexes described take space that is $\mathcal{O}(a_r n_o + n_i)$, as outgoing mail information is stored multiple times, once for each recipient. By a simple case analysis, it is easy to see that all features, other than textual similarity, can be computed using our index structures, in time that is at most logarithmic in n_i or n_o. Textual similarity, which is computed using Lucene, takes time $\mathcal{O}(n_o n_t)$. However, unlike all other features, this computation is not performed once for each contact. Instead, a single search is performed for the new mail, and the derived results determine the feature value for textual similarity for all contacts.

Hence, taken together, the indexes allow for efficient learning and ranking of mail recipients.

6 Experimental Evaluation

We used three datasets to evaluate our method, as detailed in Table 4. The Gmail dataset was extracted from private mailboxes of Gmail users who participated in our experiments. Note that we did not collect the data itself, but only the required

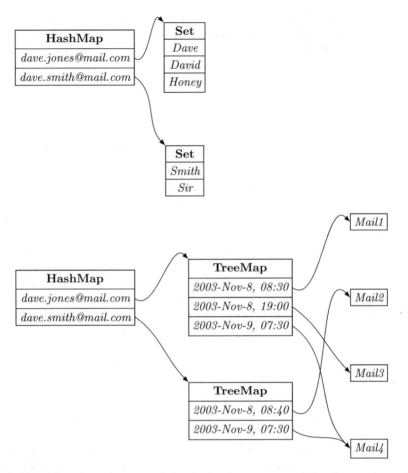

Fig. 7 Illustration of greeting names index (*top*) and a per-contact index for outgoing mails (*bottom*). Note that, in the latter, every mail can appear in multiple entries, as it can be sent to multiple recipients

feature values. The political dataset is a collection of freely available mailboxes of three well-known public figures: John "Jeb" Bush, Elena Kagan and Sarah Palin.

The predictable emails are obtained by discarding duplicates, incoming emails, the first 10 emails of a user, reply emails and emails to a new contact. In all these cases the prediction task is either meaningless or trivial.

Recall that the features "more recent outgoing percentage" and "more recent ingoing percentage" use a parameter α, whose goal was to further differentiate between the cases in which a message was sent to (received from) a contact c a long time ago, or never at all. Our experimentation uses the default value of $\alpha = 2$. Interestingly, additional experimentation with this parameter has shown that it has little affect of the quality of the result. For example, the combined MRR scores on all datasets for values of $\alpha \in \{1, 2, 4, 8, 16, 32\}$ differ one from another by at most

Table 4 Datasets

Dataset	Users	All emails	Predictable
Enron	140	517,401	42,028
Gmail	10	–	12,536
Political	3	323,180	13,309
Total	153	>840,581	67,873

Due to privacy concerns, we extracted from private Gmail mailboxes only the bare minimum necessary for our experimentation. Therefore, the number of all mails in history is missing

0.006. This seems to be because such contacts usually receive a very low ranking, both if they have not contacted c for a long time and if they have never contacted c.

6.1 Personalized Functions

In our experimentation we learned a personalized ranking function for each user in each dataset. For each user, we used the first half of the predictable mails as a training set and the remainder as a test set. We tested our prediction in two settings:

- **Full Ranking:** Rank all contacts.
- **Known-Character Ranking:** Rank only contacts with addresses beginning with the same character as some correct recipient. This setting simulates a scenario where the user correctly inserts the first character of a recipient and the system attempts to "auto-complete" the recipient's address.

Tables 5 and 6 contain the MRR (mean reciprocal rank) value for our personalized functions (Line 1), for functions determined by a single feature (Lines 2–8) and for the baseline random guess (Line 9), for full ranking and for known character ranking. As seen in bold in this figure, our personalized functions outperformed all ranking functions induced by each feature alone, in almost every case.[3] Comparison with the baseline "random guess" demonstrates the effectiveness of both the features and the personalized functions.

---

[3]In one case "more recent incoming percentage" outperformed the personalized function. This seems to be due to the fact that the Gmail datasets were mostly small accounts, except for two very large accounts of users—including one of the authors of this article—who have a compulsive habit of immediately answering every email they receive. For larger and more diverse datasets, we expect the personalized function to be the best performing.

Table 5 MRR scores of our personalized functions in comparison with ranking functions induced by each feature alone for full ranking

Method	Full ranking			
	Overall	Enron	Gmail	Political
Personalized function	**0.470**	**0.468**	**0.437**	**0.526**
More recent outgoing percentage	0.326	0.303	0.405	0.329
Outgoing textual similarity	0.322	0.327	0.257	0.397
More recent incoming percentage	0.274	0.202	**0.491**	0.312
Outgoing message percentage	0.272	0.283	0.186	0.350
Incoming textual similarity	0.213	0.206	0.192	0.278
Incoming message percentage	0.199	0.193	0.154	0.299
Greeting	0.112	0.142	0.056	0.043
Random guess	0.038	0.044	0.021	0.026

Table 6 MRR scores of our personalized functions in comparison with ranking functions induced by each feature alone for known character ranking

Method	Known-character ranking			
	Overall	Enron	Gmail	Political
Personalized function	**0.780**	**0.751**	**0.813**	**0.838**
More recent outgoing percentage	0.701	0.667	0.768	0.743
Outgoing textual similarity	0.588	0.604	0.576	0.553
More recent incoming percentage	0.564	0.521	0.807	0.474
Outgoing message percentage	0.606	0.589	0.579	0.687
Incoming textual similarity	0.415	0.404	0.504	0.366
Incoming message percentage	0.491	0.489	0.560	0.435
Greeting	0.276	0.327	0.257	0.133
Random guess	0.194	0.224	0.190	0.113

Figure 8 shows the probability (averaged over all datasets) to have at least one correct recipient in the top-*k* ranked contacts in three settings: random guess for full ranking, personalized function for full ranking and personalized function for known-character ranking. The effectiveness of our system is immediately apparent.

Our personalized function performed especially well on the political dataset. This is probably due to the fact that the mailboxes for the political users were particularly large, and hence, our function could learn with greater accuracy. Nevertheless, the personalized function performed well on all corpora. The results also show that our feature set was sufficiently small and indicative to be accurately learned even when just a small training set is available (small mailboxes are common in the Enron and Gmail datasets).

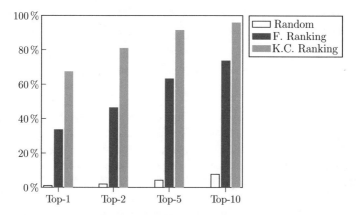

Fig. 8 Probability to have a hit among the top-k ranked elements

6.2 Within-Domain Setting

When attempting to predict the behaviour of new users, there is a very limited data generated by the user. As this data might not be sufficient for training an accurate personalized function, we would like to explore some alternatives. One alternative is training a function on a group of similar users. Similarity between users is a complex notion, but we will simplify it. We will assume that all users in the same dataset (Enron, Gmail or political) are similar users. This assumption seems reasonable to some extent. For example, users belonging to the same corporation (i.e. Enron) are likely to show similar patterns in their email behaviour. We will call a function that was trained on such group of similar users a *domain function*.

Domain functions can be considered for predicting new users, but not only for that purpose. Training on a group consisting of multiple users can rely on much larger training set than personalized function. This advantage can "compensate" for the difference between users' behaviour. Furthermore, a domain function that is trained on a large group of users can learn the group's patterns and neglect patterns that are specific for only few users within the group.

We used a separate test setting for our experimental evaluation of domain functions. In this setting, each dataset was split into two groups of users, a training group and a testing group. We used each training group to learn a domain function, yielding three domain functions, as detailed in the second row of Table 7. For testing sets we used the second half of mails of each user in the testing group. Although we could have used *all* mails of each user to test the domain function alone, we had to omit some mails to be able to train a personalized function for each user in the testing group. In this way, we could compare the personalized functions with the domain function on precisely the same test set.

An additional test setting for domain functions was the *scaled domain function*. Note that the training data for domain functions was significantly larger than that of personalized functions. Thus, to measure the quality of the training data itself,

Table 7 Comparison
between MRR scores of
personalized functions and
domain functions

Function	Enron	Gmail	Political
Personalized function	0.470	0.466	0.554
Domain function	0.481	0.471	0.554
Scaled domain function	0.456	0.464	0.539

another setting was added. In this setting, for each user, a subset of the domain training data was randomly selected. The size of the subset was equal to the size of the personal training data of the user. A scaled domain function was then trained only on the subset of the domain's training data, and tested on the user's test set. The averaged results for scaled domain functions are presented in the third row of Table 7.

As expected, personalized functions outperformed scaled domain functions on all domains. Obviously, using training data generated by the same user will allow the system to learn the individual patterns of the user and will thus result in a more accurate prediction function. Regular domain functions, on the other hand, performed slightly better than personalized functions in two domains. This shows that when having a domain function trained on sufficiently large domain training set, it is a good alternative for personalized functions. Furthermore, a combination of the two approaches might be even more accurate and stable as every approach performs similarly well on its own.

6.3 Cross-Domain Setting

The domain functions discussed in Sect. 6.2 are not always available. For instance, we may wish to predict recipients for a new user, even if we do not have any training data from similar users. Even if we have training data from other users, we may not know which users are similar to the user we are interested in. For such cases we introduce a new type of function, a *cross-domain function*. A cross-domain function is a function learned from a group of users, just like a domain function. Yet, a cross-domain function can be learned from any group of users and be applied on any other group of users. These two groups of users do not have to belong to the same domain. This implies that we can learn one single cross-domain ranking function and apply it to any new user.

The cross-domain test setting, used in the experimentation, was very similar to the in-domain setting. The difference was that in this setting every domain function was tested also on the two other domains. Comparing with the in-domain functions which reside on the main diagonal of Table 8, we can measure the cross-domain adaptivity of our model. Observe that in a few cross-domain settings there was a significant drop in performance compared to the in domain setting. This happened when training on Enron and testing on the political dataset. Also, a smaller drop in performance was observed when training on Enron and testing on Gmail, and even

Table 8 Cross-domain
results

Function	Enron	Gmail	Political
Enron function	0.481	0.428	0.441
Gmail function	0.491	0.471	0.557
Political function	0.487	0.444	0.554

Each row shows the score of the respective function on every domain

smaller drop when training on the political dataset and testing on Gmail. The other
three cross-domain settings yielded similar and even slightly better results relative
to the respective in-domain setting.

6.4 Known Prefix

In this section we perform an additional evaluation of the known prefix scenario on
two of the datasets. In Sect. 6.1 we have seen how the performance is affected when
the system is given the first character of some recipient's address. Now we wish
to extend this scenario to the case where prefixes of larger length are given to the
system. We use Enron and the political datasets for this evaluation. Our goal is to
measure the improvement in the system's performance as more and more characters
of the recipient's address are inserted by the user.

Note that a known prefix scenario is somewhat different than the basic scenario
(full ranking) when considering multiple recipients. In the basic scenario the system
attempts to predict any recipient, as long as the user is interested in adding more
recipients. That is, at every moment the system will output the top-ranked contact
that has not been outputted so far. However, in the known prefix scenario, the system
eliminates all contacts that do not begin with the given prefix. When it comes
to multiple recipients, in some cases, if not most cases, the system will end up
eliminating some actual recipients. Hence, it is possible that sometimes the system
will be better off predicting a very high ranked contact that does not begin with the
given prefix. Typically, in such cases there will be a large set of contacts beginning
with the given prefix and they will all have low or similar ranking. Yet, we claim
that using the elimination rule will increase the performance in the vast majority of
cases. The enormous decrease in the amount of contacts remaining, together with
the assurance that at least one of them is a correct recipient, compensates for the
possibility of eliminating high-ranked recipients.

Figure 9 shows the MRR score on the Enron dataset as a function of the number
of known prefix characters. When two characters are inserted, the MRR of the
system is around 0.9, and reaches 0.99 at eight characters. The results strengthen
the assumption that eliminating all contacts with other prefixes will vastly improve
the performance.

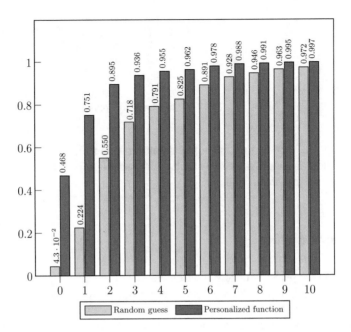

Fig. 9 MRR score on the Enron dataset as a function of the number of known prefix characters

Similar results for the political dataset are shown in Fig. 10. Note that in this data set there are significantly more contacts on average as the random guess scores are lower. Nevertheless, the scores for the personalized functions are very similar to those observed in the Enron dataset.

Finally, Fig. 11 shows the probability to have a hit in the top-k on the Enron dataset. Given two characters, the system will be correct in the first guess in around 83% of the times. Given an additional character, the first guess will be correct with probability of 89%.

6.5 Feature Evaluation

In this test setting, the contribution of each feature was measured. For each feature, a model where this feature is omitted was tested on all datasets. That is, one model was trained on every combination of six features out of the seven. Obviously, a large drop in performance means that the omitted feature is very essential for the model. That means that it has a strong correlation with the prediction and small overlap with other features.

The general setting used for this evaluation was full ranking with personalized functions. The scores were averaged across mails from all datasets.

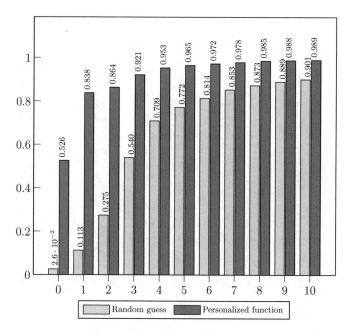

Fig. 10 MRR score on the political dataset as a function of the number of known prefix characters

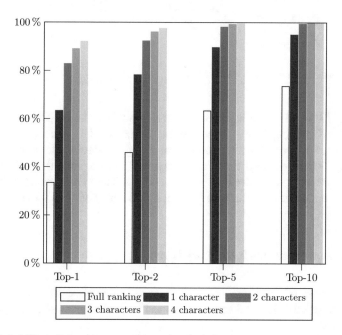

Fig. 11 Probability to have a hit among the top-*k* ranked elements

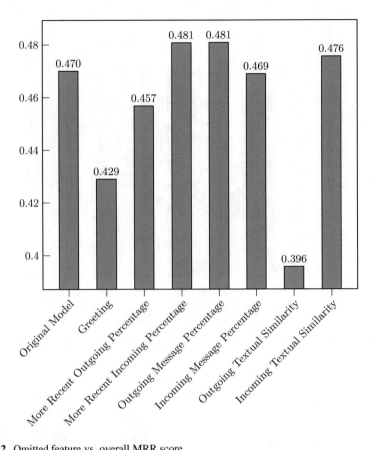

Fig. 12 Omitted feature vs. overall MRR score

In Fig. 12 the MRR score of each of these models is shown. The most essential features are Outgoing Textual Similarity and Greeting. The omission of a few of the features, such as More Recent Incoming Percentage, Outgoing Message Percentage, Incoming Textual Similarity, slightly improve the score. Note that this behaviour was not observed on the development set and it seems to have small significance. Frequency and recency features can have a significant overlap in some domains.

6.6 Comparison with Related Work

In this section we compare our system to systems presented in [11–14]. We reproduced the training and testing sets used in [14], and show a comparison of the results.

Table 9 Comparison of
MRR scores with the related
work

	Standard	Threaded
GTD	0.524	0.582
VC	0.440	0.516

There were several differences between the test setting used in [14], and the test setting used in this article for the Enron dataset. First, the evaluation in [14] was done only on 36 of the Enron users. Second, reply mails were included in the test set. Third, for most of the users, only a small fraction of the mails were assigned to the test set. Recall that in our experimentation, half of the mails of each user were assigned to the test set.

We show the MRR scores of both works in Table 9. We denote the method presented in [14] by *VC* and the method presented in this article by *GTD* (for *greeting*, *text* and *date*, the main features used in our system). A comparison of the basic methods is shown in the first column of Table 9.

The second column compares slightly modified models. Since reply mails are included in this setting, having thread information can help predicting the recipient. The threaded setting adds a wrapping rule to the model. This rule ranks all contacts appearing in previous mails of the given thread on top, regardless of the model's ranking. Note that Enron dataset does not contain direct thread information. Thus, mails were considered to belong to the same thread if their subject was the same (after omission of RE: and FWD: tags). Clearly, this method can dramatically improve the prediction's performance on reply mails. Since [14] reported an improved score achieved by wrapping their model with this rule, we also used this rule to wrap our model. The respective MRR scores are compared in the second column of Table 9.

The comparison shows that our method achieves significantly better performance. This is due to our novel greeting feature, and perhaps due to various differences in the temporal and text similarity features. We also believe that our features are more adaptive between different users. Unfortunately, [14] did not perform cross-user experimentation.

7 Conclusion and Future Work

In this article we addressed the ranking problem for email recipients. We defined a small set of features that highly correlate with recipients of a given mail. In our feature set we also defined a novel greeting feature, that proved to be beneficial for recipient prediction. In addition, we combined these features in a learning scheme, resulting with ranking functions that outperform every feature. Our learning scheme also outperformed a state-of-the-art method.

We performed experimental evaluation on two datasets, in addition to the standard Enron dataset. These two datasets were collected only after the model was constructed. Therefore, the performance on these datasets can show good adaptivity of our model. Such adaptivity was not demonstrated for previous work.

Another contribution of this work is the cross-user approach. This approach was not tested in previous work. We have shown that using our scheme for learning from different users, we can achieve high accuracy in this prediction task. This capability can be used to predict recipients for new users. It may even be used to enhance the accuracy for all users.

We note that this work was focused on the feature selection problem. For the learning itself, we used one well-known algorithm with default parameters. This implies that a more extensive study of learning methods for this task can yield an even more accurate model. In particular, online learning frameworks, such as those in [19], may fit to this task even better. Observe that after an email is sent, the recipients of the email can be given to the model. The model can then continuously tune itself and learn from its mistakes.

Overall, we conclude that in most cases, it is possible to successfully predict recipients. We also claimed, and demonstrated in [17], that this capability has various practical usages. In [17], we demonstrated a recipient auto-complete mechanism, as well as an alerting system that is triggered when attempting to send mails to unlikely recipient. Another potential application is to suggest adding highly ranked contacts to the mail. This application can improve email communication and effectiveness, since many users, especially in big corporations, have difficulties recalling all relevant recipients of an email. We also note that our model enables learning certain aspects of the user's behaviour, just like [3], and therefore can be also used to detect abnormal email activity, and form a defense against the spread of worms and spam. We believe that in the future, recipient prediction systems will be added to most popular email clients. According to the results shown in this work, we deduce that such systems are likely to improve email communication, user experience, and provide a credible safety net against email leaks. Increased usage of such systems can also help in their evaluation and provide more means for training improved models.

Acknowledgements Zvi Sofershtein and Sara Cohen were partially supported by the Israel Science Foundation (Grant 1467/13) and the Ministry of Science and Technology (Grant 3-9617).

References

1. Radicati S, Levenstein J. Email statistics report, 2015–2019. Technical report, The Radicati Group; 2015.
2. Shen J, Brdiczka O, Liu J. Understanding email writers: personality prediction from email messages. In: Carberry S, Weibelzahl S, Micarelli A, Semeraro G, editors. User modeling, adaptation, and personalization. Lecture notes in computer science, vol. 7899. Berlin, Heidelberg: Springer; 2013. p. 318–30.
3. Dabbish LA, Kraut RE, Fussell S, Kiesler S. Understanding email use: predicting action on a message. In: Proceedings of the SIGCHI conference on human factors in computing systems, CHI '05. New York: ACM; 2005. p. 691–700.

4. Aberdeen D, Pacovsky O, Slater A. The learning behind Gmail Priority Inbox. In: NIPS 2010 workshop on learning on cores, clusters and clouds; 2010.
5. Ayodele T, Zhou S, Khusainov R. Email reply prediction: a machine learning approach. In: Salvendy G, Smith MJ, editors. Human interface and the management of information. information and interaction. Lecture notes in computer science, vol. 5618. Berlin, Heidelberg: Springer; 2009. p. 114–23.
6. Karagiannis T, Vojnovic M. Behavioral profiles for advanced email features. In: Quemada J, León G, Maarek YS, Nejdl W, editors. WWW. New York: ACM; 2009. p. 711–20.
7. Martin S, Sewani A, Nelson B, Chen K, Joseph AD. Analyzing behaviorial features for email classification. Berkeley, CA: University of California; 2005.
8. Dredze M. "Sorry, i forgot the attachment:" email attachment prediction. In: Proceedings of the third conference on E mail and anti spam (CEAS); 2006.
9. Ghiglieri M, Fürnkranz J. Learning to recognize missing e-mail attachments. Technical report TUD-KE-2009-05. Knowledge Engineering Group. Darmstadt: TU Darmstadt; 2009.
10. Shetty J, Adibi J. Enron email dataset. Technical report. Marina Del Rey, CA: USC Information Sciences Institute; 2004.
11. Carvalho VR, Cohen WW. Ranking users for intelligent message addressing. In: Proceedings of the IR research, 30th European conference on advances in information retrieval, ECIR'08. Berlin, Heidelberg: Springer; 2008. p. 321–33.
12. Carvalho VR, Cohen WW. Preventing information leaks in email. In: Proceedings of SIAM international conference on data mining (SDM-07), Minneapolis, MN; 2007.
13. Carvalho VR. Modeling intention in email - speech acts, information leaks and recommenda-tion models. Studies in computational intelligence, vol. 349. Berlin: Springer; 2011.
14. Carvalho VR. Modeling intention in email. Ph.D. thesis. School of Computer Science Carnegie Mellon University, Pittsburgh, PA, 2008.
15. Roth M, Ben-David A, Deutscher D, Flysher G, Horn I, Leichtberg A, Leiser N, Matias Y, Merom R. Suggesting friends using the implicit social graph. In: Proceedings of the 16th ACM SIGKDD conference on knowledge discovery and data mining; 2010.
16. Bartel JW, Dewan P. Towards hierarchical email recipient prediction. In: CollaborateCom; 2012. p. 50–9.
17. Sofershtein Z, Cohen S. Predicting email recipients. In: Proceedings of 2015 IEEE/ACM international conference on advances in social networks analysis and mining; 2015.
18. Joachims T. Training linear SVMs in linear time. In: Proceedings of the 12th ACM SIGKDD international conference on knowledge discovery and data mining, KDD '06. New York: ACM; 2006. p. 217–26.
19. Fiat A, Woeginger G, editors. Online algorithms: the state of the art. Lecture notes in computer science. New York: Springer; 1998.

A Credibility Assessment Model for Online Social Network Content

Majed Alrubaian, Muhammad Al-Qurishi, Mabrook Al-Rakhami, and Atif Alamri

1 Introduction

The Internet has evolved into a tool for global communication. It has revolutionized the way millions of users communicate through online social networks (OSNs) such as Twitter, YouTube, Facebook, Snapchat, LinkedIn, and Keek. These social networks have become extremely popular in recent years [1, 2]. As an integral part of daily life, they enable people to share their interests, follow the latest trends, and communicate with friends. On the other hand, concerns over credibility on the Internet have grown, especially on OSNs. Information is regarded as a type of currency in today's digital age, and information publishing is now widely perceived as a right. As such, many studies have investigated Web credibility since the emergence of social networks as a social phenomenon [3].

Credibility refers to "the quality of being convincing or believable" or "the quality of being trusted and believed in." It is also defined as "trustworthiness, believability, reliability, accuracy, fairness, objectivity, and dozens of other concepts and combination thereof." Furthermore, credibility has been defined "in terms of characteristics of persuasive sources, characteristics of the message structure and content, and perceptions of media" [4]. It has attracted considerable research attention in many fields, including information science, management, finance, health, psychology, sociology, political science, human–computer interaction (HCI), communication, and information retrieval (IR).

M. Alrubaian • M. Al-Qurishi (✉) • M. Al-Rakhami • A. Alamri
College of Computer and Information Sciences, Research Chair of Pervasive and Mobile
Computing, King Saud University, Riyadh, Saudi Arabia
e-mail: malrubaian.c@ksu.edu.sa; qurishi@ksu.edu.sa; malrakhami@ksu.edu.sa; atif@ksu.edu.sa

© Springer International Publishing AG 2017
M. Kaya et al. (eds.), *From Social Data Mining and Analysis to Prediction
and Community Detection*, Lecture Notes in Social Networks,
DOI 10.1007/978-3-319-51367-6_3

The proliferation of tweets has made it extremely difficult for users to assess the credibility of information on Twitter, a popular microblogging Web site. This problem has prompted the development of algorithms and tools for filtering tweets. Most solutions to this problem have been developed on the basis of definitions, pre-suppositions, and approaches in specific fields.

In terms of OSNs, credibility has been studied from several perspectives without definitive results. For example, studies on information trust, recommendation, and reputation have evaluated the impact of social networks. In recent years, concerns over information credibility on ONSs have grown, especially during high-impact events such as the Boston Marathon bombings, Hurricane Sandy, and the Arab Spring. The use of OSNs during these events has created opportunities for information diffusion that would otherwise not exist.

Credibility in the online domain has been studied extensively. Some notable studies on credibility are based on various factors such as newsworthiness, content credibility, source credibility, doctored images, and rumors [5]. Accordingly, developing a model for identifying "true" information on Twitter is a challenging task. In this regard, several researchers have proposed methods for computing the credibility of tweets. Some of them have adopted machine-learning approaches [4, 6, 7], whereas others have adopted approaches based on human perception and judgment [8, 9]. However, no clear conclusion or consensus has been reached in terms of what can be done to ascertain credibility. Therefore, further specialized studies are required in this regard.

This paper goes a step further by improving upon previously developed methods in terms of overcoming the shortcomings of their design aspects. One aspect that has not been captured in previous studies is the fact that trust, credibility, and their connoted sub-factors are subjective. In particular, we propose a credibility analysis model for identifying implausible content on Twitter in order to prevent the proliferation of false or malicious information.

Our major contributions and the key properties of the proposed technique can be summarized as follows:

- Proposal of a new credibility assessment model that maintains complete entity-awareness (tweet, user) to reach a precise information credibility judgment: This model consists of six integrated components that operate in an algorithmic form to analyze and assess the credibility of tweets.
- Enhancement of our classifier by weighting each feature according to its relative importance: This weighting methodology implements a pair-wise comparison to produce a priority vector that ranks the features of an instance according to their relative importance as well as with respect to the user's need and the investigated topic.
- Validation of our model by applying it to two large datasets of Twitter content with and without the relative importance algorithm: Our results show that the model with the relative importance approach provides reasonably accurate credibility assessment.

The remainder of this paper is organized as follows. Section 2 reviews related studies on credibility assessment. Section 3 describes the proposed model and its six major components in detail. Section 4 presents our experimental results. Finally, Sect. 5 summarizes our findings and concludes the paper by briefly discussing the directions for future work.

2 Related Work

Credibility on OSNs has attracted considerable research interest from various perspectives. In general, credibility in the online domain has been investigated extensively. Several credibility research trends have been highlighted in this area, such as those involving automated and human-based approaches as well as the hybridization between them.

The literature includes a large number of studies on automated approaches based on machine learning, specifically, the supervised learning approach [1, 4]. Supervised learning techniques include decision trees, support vector machines (SVMs), and Bayesian algorithms. Castillo et al. [7] were the first group to study Twitter credibility. On the basis of various features, they proved that SVM, J48 decision tree, and naive Bayes classifiers are effective tools for assessing news topic credibility. They correctly classified instances at the topic level with an accuracy of 89.12% and realized credibility classification with an accuracy of nearly 86%.

Most studies [10–13] in this domain follow the feature extraction and classification methods of Castillo et al. [7] with slight variations. However, all of them have ignored the relative importance of features, which could significantly affect the final judgment. The credibility of information about a specific topic/event can be assessed by ranking tweets using content and user features. Gupta and Kumaraguru [14, 15] used supervised machine learning with relevance feedback to rank tweets based on their features. Through an empirical study, they showed that their approach gives good results in the case of crisis events.

Some researchers [13] have suggested that the number of friends and followers is often linked to user popularity. These features reflect highly relevant measures that can be used to determine the user's credibility. Consequently, complete awareness of an entity and its network relation attributes yields results that are more accurate than those of the credibility classification method. Here, awareness includes consideration of popularity and credible content. Thus far, only one study has investigated this issue using sentiment analysis [16], which we will use in a component of our model.

In order to implement an unsupervised approach, Al-Khalifa and Al-Eidan [17] proposed an evidence-based method that considers the similarity between tweeted news and verified content. However, this method could lead to some problems. First, it requires semantic analysis to obtain good results; otherwise, it is ineffective. Second, it will give incorrect computation results if the topics at hand are not found in trusted news sources. Third, it cannot handle media content such as videos and images.

On the other hand, several studies have analyzed credibility using human-based approaches. Post-based features affect credibility assessment as evaluators might be influenced by the context of a specific topic or event as well as political orientation. Morris et al. [8] conducted a survey to understand users' perceptions by asking participants to identify indicators of credibility on Twitter. They found that users often considered features that are easily visible, such as user name and user image. However, they also found that such features are poor indicators of credibility.

Furthermore, they showed that the features considered by a user as indicators of credibility are different from those used by Web search tools such as the Google search engine. Schmierbach and Oeldorf-Hirsch [18] showed that the credibility of user-evaluated news items posted on Twitter is significantly lower than that of new items on traditional media Web sites. Morris et al. [9] found that science-related tweets had the highest level of perceived credibility. Thus, disregarded such tweets might skew the overall credibility rating. Yang et al. [19] conducted an online study of user perspectives on two microblogging services, Twitter in the USA and Weibo in China, in order to compare the impact of several factors on both networks, including gender, name style, profile image, location, network overlap, and message topic. The survey data showed a key variance between the two countries. In addition, the study showed that users in China are more inclined to trust Weibo as an information source than American users of Twitter.

Most human-based approaches suffer from two main drawbacks: inconsistency and low reliability. To guarantee an evaluator's reliability, some studies [20] have proposed two gold standard questions. If an evaluator's answers to these questions are correct, then his/her judgment will be acceptable; otherwise, he/she will be rejected. This could be a problem if most of the evaluators' answers are incorrect. By reviewing more than 107 papers in this field, we found that machine-learning classification techniques provide more accurate results than human-based approaches.

Only a few studies have investigated hybrid methods that combine the advantages of both automated and human-based approaches to analyze social media credibility [1, 21, 22]. For instance, supervised approaches, such as the naive Bayes method, have been combined with human perception surveys [22], while clustering approaches have been combined with weighting and information retrieval algorithms [23].

Furthermore, some researchers have built systems to assess the credibility of tweets. Truthy [24, 25] is one such system that conglomerates, analyzes, and visually represents information in the domain of trending political topics. A system that analyzes the credibility of tweets in Arabic has been developed by Al-Eidan et al. [26]. In addition, a Web-based application—TweetCred—has been developed to compute the credibility scores of tweets using a supervised ranking algorithm [27].

3 Proposed Model

This section describes the six components of the proposed model: (1) tweet collector, (2) tweet feature extractor, (3) features relative importance matrix, (4) sentiment score process, (5) classification process, and (6) final credibility assessment. The architecture of the proposed model is shown in Fig. 1. Each component is explained in detail in the following subsections. In principle, all six components together represent an iterative process that combines an automated approach and a human-based approach to yield better credibility assessment results with high accuracy.

3.1 Tweet Collector

This component uses the Twitter API as a source of information for whole datasets. Both the Twitter streaming API and the Twitter search API are used to collect tweets regarding different topics. The former API was used to collect datasets about ISIS using the keywords listed in Table 1; these keywords were selected as they usually refer to ISIS. The latter API was used to collect the 3200 most recent user tweets as allowed by Twitter.

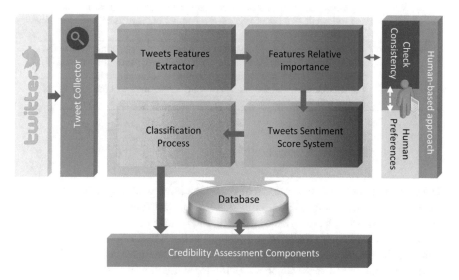

Fig. 1 Architecture of the proposed model

Table 1 Arabic and English keywords used in data collector

Arabic keyword	English translation
داعش	ISIS
دولة العراق والشام	The State of Iraq and the Levant
حسابات داعش	ISIS accounts
صواريخ داعش	ISIS missiles
دبابات داعش	ISIS tanks
حدود السعودية داعش	Saudi Arabia borders ISIS
داعش العراق	ISIS Iraq
احتلال داعش	Occupation ISIS
تفجيرات داعش	ISIS bombings, blasts
داعش سوريا	ISIS Syria

Table 2 Feature level extraction

Feature level	Description
Tweet level	Text features: characteristics related to the content of the tweet, such as the length of the message, number of replies, and/or number of retweets, which may reflect the importance of the tweet. In addition, tweets containing hashtags and "@" mentions as well as URLs and static/animated emoticons Sentiment features: number of positive and negative words, based on a predefined list of sentiment words
User level	Some of these features are latent and some of them are explicitly revealed in user profiles. For example, age, gender, education, political orientation, and even user preferences are considered as latent attributes. The number of followers, number of friends, and number of retweets as well as replies to a user's tweets
Topic level	Extracting topic-based features is the process of aggregating most of the tweet-based features, such as the URL fraction of tweets, the hashtag (#) fraction of tweets, and the average sentiment score of tweets. The number of duplications, i.e., the number of times a user may post the same tweet more than once

3.2 Tweet Feature Extractor

The collected content is in the form of semi-structured data, e.g., JSON files, with attributes such as status, URLs, embedded media content (i.e., images/videos), number of followers, friends, lists, hashtags, and mentions. To assess the credibility of Twitter content, we extracted tweet features at three levels: post (message) level, topic (Event) level, and user level. For each level, we used aggregated features such as the number of retweets on a specific topic. Each user has a personal record of information for his/her profile. We omitted those users who have no followers. Then, we performed credibility assessment for the three levels. Several researchers have opted to use a hybrid credibility measurement that combines the three above-mentioned levels. Table 2 summarizes the three levels of feature extraction.

All these refined and extracted features are stored in MySQL DB. We consider hybrid-level feature extraction to perform our credibility assessment for the following reasons. First, at the tweet level, the 140-character limit of Twitter messages makes them inappropriate (to a certain degree) for analysis with high-impact topic models. From another perspective, individual tweets do not provide sufficient information about a combination of precise latent topics. Second, users can quickly and easily earn thousands of followers through Twitter follower marketing.

3.3 Relative Importance Matrix

We believe that the extracted features should be weighted before assessing a given tweet, user, or topic because of the importance of the features that affect the final judgment of credibility. By investigating more than 112 papers on the credibility of social Web content, we found that the number of followers was considered as the most important feature (68 papers), followed by message URLs (63 papers). On the other hand, time zone (4 papers), media (2 papers), and number of favorites (1 paper) were considered as the least important features, as shown in Fig. 2. Therefore, the importance of the used features has a significant influence on the classification process. Not all features are quantitative; some of them are qualitative and require human intervention to determine their importance with respect to the overall goal. Such intervention happens only once. In this regard, we consider an expert (i.e., a human) to generate a judgment matrix regarding the importance of each feature.

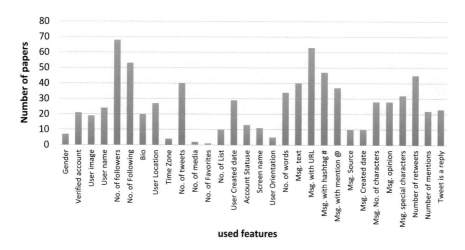

Fig. 2 Most important features considered in the literature

Table 3 Numerical scale for pair-wise comparisons

Scale	Description
1	Equal importance: two features contribute equally to the objective
3	Moderate importance: slightly favors one over another
5	Essential or strong importance: strongly favors one over another
7	Demonstrated importance: dominance of one demonstrated in practice
9	Extreme importance: evidence favoring one over another of higher possible order of affirmative
2, 4, 6, 8	Intermediate importance: when a compromise is needed

To determine the relative importance of the extracted features/attributes, we used a pair-wise comparison method, which is based on the analytical hierarchy process proposed by Saaty [28]. This comparison process generates a measurement scale of relative priorities or weights for the extracted features. The pair-wise comparisons are made by ratios of importance (when comparing features/attributes), which are evaluated on a numerical scale from 1 to 9, as shown in Table 3.

The pair-wise comparisons are performed at the attribute level, where they are compared, two at a time, with respect to the overall goal of the credibility decision process. The results of the pair-wise comparisons are entered into a matrix. An activity is equally important when compared to itself; thus, the diagonal elements of the matrix must be "1". In addition, the matrix must satisfy the relation $a_{ij} = \frac{1}{a_{ij}}$. The matrix is read as follows: for example, if $a_{ij} = 5$, then activity i is of essential or strong importance over activity j. This also implies that the value of a_{ij} will be equal to 1/5. This matrix is the basic unit of the analysis and is called the pair-wise comparison matrix or the judgment matrix.

After all the pair-wise comparison values are entered into the matrix, a vector of priorities is computed for the attributes. In mathematical terms, the principal eigenvector (eigenvector associated with the maximum eigenvalue) is obtained, and when it is normalized by dividing each element of the matrix by the sum of its column elements, it becomes the vector of priorities. Algorithm 1 outlines the computation of the priority vector and the measurement of matrix consistency. Inconsistencies can occur in the matrix for various reasons, including lack of information, clerical errors, and lack of concentration. Given a set of attributes, $a = a_1, a_2, ..., a_n$, where n is the number of features, the analyst repeatedly compares one feature to another until all possible pair-wise comparisons are completed.

Algorithm 1: Compute the Priority Vector of Features

Input:X_A (note that this matrix may be inconsistent)

    Input: $RI = \{0, 0, 0.58, 0.9, 1.12, 1.24, 132, 1.41, 1.45, 1.49\}$ Random Consistency Index

Output: List of ranking attributes $\overrightarrow{RA}$ and consistency ratio CR

For each column $\overrightarrow{C}$ **In**X_A

    $\overrightarrow{Sum} \leftarrow \sum_{a_i \in \overrightarrow{C}} (a_i)$ w.r.t the row dimension

End For

For each attributea_i **In** X_A

    $NX_A \leftarrow$ **Create** Normalized Matrix by dividing each entry by the $\overrightarrow{Sum}$ of the column x

    $\overrightarrow{PV} \leftarrow$ **Calculate** Geometric Mean (NX_A) w.r.t the row dimension

End For

$\overrightarrow{RA} \leftarrow$ **Create** a list for ranking user attributes w.r.t $\overrightarrow{PV}$

  // to check the consistency

$\overrightarrow{W} \leftarrow \overrightarrow{PV} \times X_A$

$m = Sizeof\left(\overrightarrow{W}\right)$

    For each attributea_i **In**$\overrightarrow{W}$

        For each attributea_j **In**$\overrightarrow{PV}$

          If$i == j$**then**

            $\overrightarrow{Sum} \leftarrow \sum_{i \in \overrightarrow{W},\ j \in \overrightarrow{PV}} \mathbf{(ai/aj)}$

          End If

        End For

    End For

    $\lambda_{max} = Average\left(\overrightarrow{Sum}\right)$ where λ_{max}*should be close to* m

$CI = \frac{\lambda_{max} - m}{m - 1}$

    $CR = \frac{CI}{RI}$, for each CR**set threshold**θ**, where**$\theta = 0.05\ for\ m = 3$, $\theta = 0.08\ for\ m \geq 5$

    If$\theta > 0.1$ **then**

        X_A is inconsistent and should be revised

Else

        X_A inconsistency is acceptable

End If

3.4 Classification Process

We use the well-known naive Bayes classifier along with the relative importance of each feature in order to categorize the collected tweets. We have two classes: credible and non-credible. To determine the class with which a tweet T is most associated, we have to calculate the probability that tweet T is in class C_x, written as $P(C_x|T)$, where x is credible or non-credible. Using the naive Bayes method, we calculate $P(C_x|T)$ as

$$P(C_x|T) = \frac{P(T|C_x) \times P(C_x)}{P(T)} \quad (1)$$

As we are interested in relative (not absolute) values, $P(T)$ can be safely ignored because it is always constant:

$$P(C_x|T) = P(T|C_x) \times P(C_x) \quad (2)$$

Tweet T is split into a set of attributes $A = \{a_1, a_2, \ldots, a_n\}$. The probability of each attribute is multiplied by its priority vector (PV) (see Sect. 3.3) as follows:

$$P(a_i) = P(a_i|C_x) \times PV_{a_i} \quad (3)$$

Thus, with the assumption that the attributes appear independently, $P(T|C_x)$ can be calculated in the following form for each class:

$$P(a_1, a_2, \ldots, a_n|C_x) = \prod_{i=1}^{n} P(a_i) \quad (4)$$

Then, the largest value is selected to determine the class for tweet T as follows:

$$c_{NB} = \arg\max_{x \in C} P(x) \times \prod_{i} P(a_i) \quad (5)$$

3.5 Sentiment Score Process

User sentimentality has a significant influence on tweet credibility judgment with regard to an event or topic, especially when a user is oriented toward some sect or group. Some users might deliberately disseminate information that is misleading in order to create chaos, as in the case of the Arab Spring in 2011. Here, sentiments define the factors that affect social relationships, the psychological state of users, and their orientation. Sentiment analysis also involves why a trustor should trust a trustee. In one study [16], the numbers of positive and negative words have been calculated on the basis of a predefined sentiment word list. We adopted this concept of calculating the sentiment of tweets for each involved user. Checking the sentiment of each tweet of a particular user is an important step. It enables us to measure user behavior, which leads to high credibility precision. Further, it has been found that most incredible events are negative social events that generate strong negative sentiment words and opinions. To measure the sentiment of Arabic content (i.e., tweets, retweets, and replies), we used the SAMAR technique [29].

Therefore, the sentiment of user tweets should satisfy the following equations (note that we omit neutral tweets):

$$U_{\text{Pos}(T)} = \frac{\sum U_T^+}{\sum U_T^+ + \sum U_T^-} \tag{6}$$

and

$$U_{\text{Neg}(T)} = 1 - U_{\text{Pos}(T)} \tag{7}$$

3.6 Final Credibility Assessment

Based on the above-mentioned points, a metric $\text{Cred}(T)$ is defined as a simple arithmetic expression that facilitates the credibility assessment of tweet information, based on a numerical value. This is the final component of the proposed framework, where the final process of credibility assessment is carried out by combining the results of the previous components as follows:

$$\text{Cred}(T) = C_{\text{Cls}(T)} + U_{\text{Pos}(T)} + \text{SS}(T) \tag{8}$$

where $C_{\text{Cls}(T)}$ is the classifier probability that a given tweet T is credible or non-credible (as calculated by Eq. (5)); $U_{\text{Pos}(T)}$ is the value of user sentiment history (as calculated by Eq. (6)); and $\text{SS}(T)$ is the sentiment score of the given tweet. We consider that the tweet is most likely credible if the value of $\text{Cred}(T)$ is more than 50% and not credible otherwise.

4 Experiments

We applied the proposed approach to real-world data from Twitter. The crawled data was obtained over three months (August–October 2014). This dataset of 2,977,682 Tweets was split into seven typical chunks: two experimental datasets and five developmental datasets. For this study, the experimental datasets consisted of 187,614 English keywords and 186,819 Arabic keywords for tweets about "ISIS (داعش.") The following subsections describe a sample of labeled data distribution and discuss the obtained results.

4.1 Experimental Data

We created two ground truth datasets consisting of 187,614 English keywords and 186,819 Arabic keywords for tweets about "ISIS (داعش)" from 155,794 unique accounts related to these datasets. After labeling the non-credible tweets, we extracted features from them. We were not able to build a social graph from the public data because Twitter's public streaming and search APIs provide access to only public tweets, and they are not socially connected. Consequently, it is not possible for us to extract social graph-based features such as local centrality and distance properties. Such expensive features are not suitable for use in the classification process, even though they are more powerful in terms of discriminating between credible and non-credible tweets. In addition, we preferred to use features that can be computed from a tweet itself in a straightforward manner. To ascertain the relevance of the content to our objective, we discarded greeting tweets and any other redundant tweets. This step is crucial to guarantee accurate results. We extracted features from our dataset as listed in Table 2. According to the object from which the features are extracted, they can be divided into three categories: tweet-level features, user-level features, and topic-level features.

From the characteristics of these features, we plotted the cumulative distribution function (CDF) of samples from each level. From Fig. 3a, we can see that credible tweets are likely to be more positive than non-credible tweets. When it comes to the feature "Number of characters per tweet," there is no significant difference between credible and non-credible tweets. This could because users who propagate non-credible tweets try to mimic the posting behavior of credible sources. Figure 3b shows the character lengths of credible and non-credible tweets.

With respect to the number of hashtags, Fig. 3c shows that credible tweets usually have more hashtags than non-credible ones. Furthermore, credible tweets are more stable in their distribution regarding a topic/event than non-credible tweets because of a large number of outliers in the non-credible tweet distribution. Around 70.33% of non-credible tweets do not have embedded hashtags, while the corresponding proportion for credible tweets is only 58.25%. In terms of user or "@" mentions, 96.11% of the credible tweets do not have an embedded user mention, whereas 90.93% of the non-credible tweets tend to have at least two user mentions in each tweet. Finally, users who propagate non-credible tweets seem to have fewer followers and more friends than those who propagate credible tweets, as shown in Fig. 3e, f. In general, an analysis of these features reveals their relative power of discriminating between credible and non-credible news on Twitter.

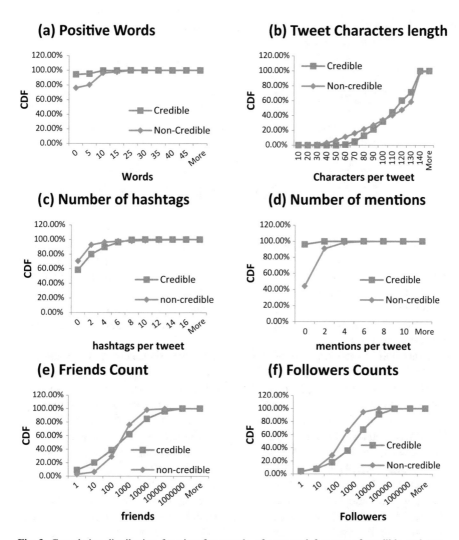

Fig. 3 Cumulative distribution function for sample of extracted features of credible and non-credible tweets

4.2 Experimental Results

In our experiment, we trained a supervised classifier to find the credibility class for a Twitter topic. In the first step of the credibility assessment, the evaluators were asked to indicate the credibility levels of tweets while justifying their judgment.

Our dataset was stored in two tables, namely Tds1 and Tds2. We used the train/test split technique to evaluate our model. This technique is preferred to other techniques in many cases owing to its speed, simplicity, and flexibility. Thus, we

Table 3 Performance of M1 credibility assessment (without relative importance algorithm)

	Precision (%)	Recall (%)	F-measure (%)	Accuracy (%)
Tds1	84.63	82.52	83.56	75.18
Tds2	86.96	85.11	86.02	79.37

Table 4 Performance of M2 credibility assessment (with relative importance algorithm)

	Precision (%)	Recall (%)	F-measure (%)	Accuracy (%)
Tds1	89.20	88.01	88.60	82.25
Tds2	90.91	89.89	90.40	85.47

split each dataset into training and testing sets. In order to evaluate the performance of the proposed model, we trained two naive Bayes models, namely M1 (with relative importance algorithm) and M2 (without relative importance algorithm). Experiments using M1 and M2 were conducted on a computer. Finally, we made a class prediction for each testing set, i.e., we applied M1 and M2 to each testing set. The output was the class prediction (credible or non-credible) for every observation (tweet) in the testing sets.

After making the predictions, we calculated the accuracy of the models on the basis of recall (also known as the *true positive rate*) or sensitivity. Recall or sensitivity is an intuitive parameter because it is a measure of how sensitive the model is to credible tweets. Maximizing the sensitivity of the classifier is important in terms of spreading news, especially during emergencies or crises. As shown in Table 4, M1 and M2 are more effective with Tds2 than with Tds1 because Tds1 has more noise in its features than Tds2, as observed during our experiments. However, M2 outperforms M1 in both Tds1 and Tds2 with recall rates of 88 and 89.89%, respectively. Another performance parameter is precision, which is a measure of how often M1 and M2 are correct when predicting a credible tweet. As shown in Tables 4 and 5, M2 yielded better precision scores than M1 for both datasets (89.2 and 90.91%). Precision and recall are very important measures that provide us with information about the underlying distribution of our testing data. Thus, the proposed model should achieve a good tradeoff between them. F-measures of 88.6% for Tds1 and 90% for Tds2 indicate that M2 achieves a good precision-recall tradeoff.

The results for models M1 and M2 are shown in Fig. 4. The average assessment accuracy was measured for each training dataset size according to the testing dataset size. Each test result was the average of three trials. We can see that the assessment accuracy increased with the use of the relative importance of the features, and it varied from one dataset to another.

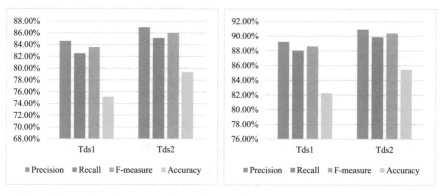

(a) M1 (without relative importance algorithm) (b) M2 (with relative importance algorithm)

Fig. 4 Performance of credibility assessment using M1 and M2

5 Conclusion

This paper proposed a novel model for credibility analysis of microblogging Web sites such as Twitter. The relative importance of the features of a tweet is a key aspect of the proposed model, which is considered together with the classification technique and the user sentiment history for tweet credibility assessment. To the best of our knowledge, this is the first study to consider the effect of relative importance on credibility assessment. Our experiments showed that relative importance has a critical effect on credibility assessment in terms of accuracy, and the results included acceptable levels of false positives or false negatives. Directions for future research include enhancement and detailed analysis of the proposed approach. In addition, we plan to modify our final credibility assessment algorithm to rank tweets in real time.

References

1. AlRubaian M, Al-Qurishi M, Al-Rakhami M, Rahman SMM, Alamri A. A multistage credibility analysis model for microblogs. Presented at: Proceedings of the 2015 IEEE/ACM International Conference on Advances in Social Networks Analysis and Mining; 2015; Paris.
2. Al-Qurishi M, Aldrees R, AlRubaian M, Al-Rakhami M, Rahman SMM, A. Alamri. A new model for classifying social media users according to their behaviors. Web Applications and Networking (WSWAN), 2015 2nd World Symposium, Tunisia: IEEE; 2015. pp. 1–5.
3. Al-Qurishi M, Al-Rakhami M, AlRubaian M, Alarifi A, Rahman SMM, Alamri A. Selecting the best open source tools for collecting and visualizing social media content. in Web Applications and Networking (WSWAN), 2015 2nd World Symposium, Tunisia: IEEE; 2015. pp. 1–6.

 4. AlMansour AA, Brankovic L, Iliopoulos CS. Evaluation of credibility assessment for microblogging: models and future directions. Proceedings of the 14th International Conference on Knowledge Technologies and Data-driven Business, 2014; New York: ACM. p. 32.
 5. AlRubaian M, Al-Qurishi M, Rahman SMM, Alamri A. A novel prevention mechanism for sybil attack in online social network. Presented at the WSWAN'2015, Tunisia: IEEE; 2015.
 6. Gayo-Avello PTM, Mustafaraj E, Strohmaier M, Schoen H, Peter Gloor D, Castillo C, Mendoza M, Poblete B. Predicting information credibility in time-sensitive social media. Internet Res. 2013;23:560–88.
 7. Castillo C, Mendoza M, and Poblete B. Information credibility on twitter. Presented at: Proceedings of the 20th international conference on World wide web; 2011; Hyderabad.
 8. Rieh SY, Morris MR, Metzger MJ, Francke H, Jeon GY. Credibility perceptions of content contributors and consumers in social media. Proc Am Soc Inf Sci Technol. 2014;51:1–4.
 9. Morris MR, Counts S, Roseway A, Hoff A, Schwarz J. Tweeting is believing?: understanding microblog credibility perceptions. Proceedings of the ACM 2012 conference on Computer Supported Cooperative Work, New York: ACM; 2012. pp. 441–50.
10. Indrawan-Santiago M, Han H, Nakawatase H, Oyama K. Evaluating credibility of interest reflection on twitter. Inter J Web Inf Syst. 2014;10:343–62.
11. Canini KR, Suh B, Pirolli PL. Finding credible information sources in social networks based on content and social structure. Privacy, Security, Risk and Trust (PASSAT) and 2011 IEEE Third International Conference on Social Computing (SocialCom); 2011. pp. 1–8.
12. Sikdar SK, Kang B, O'Donovan J, Hollerer T, Adal S. Cutting through the noise: defining ground truth in information credibility on Twitter. Human. 2013;2:151–67.
13. Saez-Trumper D. Fake tweet buster: a webtool to identify users promoting fake news ontwitter. In: Proceedings of the 25th ACM conference on Hypertext and social media. New York: ACM; 2014. P. 316–317.
14. Gupta A, Kumaraguru P. @Twitter credibility ranking of tweets on events #breakingnews. IIIT-Delhi; 2011.
15. Gupta A, Kumaraguru P. Credibility ranking of tweets during high impact events. Proceedings of the 1st Workshop on Privacy and Security in Online Social Media, 2012; New York: ACM. p. 2.
16. Ikegami Y, Kawai K, Namihira Y, Tsuruta S. Topic and opinion classification based information credibility analysis on twitter. In: Proceedings of 2013 IEEE international conference on systems, man, and cybernetics (SMC). Washington DC: IEEE Computer Society; 2013. P. 4676–81.
17. Al-Khalifa HS, Al-Eidan RM. An experimental system for measuring the credibility of news content in Twitter. Int J Web Inf Syst. 2011;7:130–51.
18. Schmierbach M, Oeldorf-Hirsch A. A little bird told me, so I didn't believe it: Twitter, credibility, and issue perceptions. Commun Q. 2012;60:317–37.
19. Yang J, Counts S, Morris MR, Hoff A. Microblog credibility perceptions: comparing the USA and China. Proceedings of the 2013 Conference on Computer Supported Cooperative Work, New York: ACM; 2013. pp. 575–86.
20. Shariff SM, Zhang X, Sanderson M. User perception of information credibility of news on twitter. In: Advances in information retrieval. Springer; 2014. P. 513–8.
21. AlMansour AA, Brankovic L, Iliopoulos CS. A model for recalibrating credibility in different contexts and languages-a twitter case study. Int J Digit Inf Wirel Commun (IJDIWC). 2014;4:53–62.
22. Kang B, O'Donovan J, Höllerer T. Modeling topic specific credibility on twitter. Proceedings of the 2012 ACM International Conference on Intelligent User Interfaces; 2012; New York: ACM. pp. 179–88.
23. Ito J, Song J, Toda H, Koike Y, Oyama S. Assessment of tweet credibility with LDA features. Proceedings of the 24th International Conference on World Wide Web Companion; 2015; New York: ACM. pp. 953–58.

24. McKelvey KR, Menczer F. Truthy: enabling the study of online social networks. Proceedings of the 2013 Conference on Computer Supported Cooperative Work Companion; 2013; New York: ACM. pp. 23–6.
25. Ratkiewicz J, Conover M, Meiss M, et al. Truthy: mapping the spread of astroturf in microblog streams. Proceedings of the 20th International Conference Companion on World Wide Web; 2011; New York: ACM. pp. 249–52.
26. Al-Eidan RMB, Al-Khalifa HS, Al-Salman AS. Towards the measurement of arabic weblogs credibility automatically. Proceedings of the 11th International Conference on Information Integration and Web-based Applications & Services; 2009; New York: ACM. pp. 618–22.
27. Gupta A, Kumaraguru P, Castillo C, Meier P. Tweetcred: real-time credibility assessment of content on twitter. In: Social informatics. Springer; 2014. P. 228–43.
28. Saaty TL. Decision making with the analytic hierarchy process. Int J Serv Sci. 2008;1:83–98.
29. Abdul-Mageed M, Diab M, Kübler S. SAMAR: subjectivity and sentiment analysis for Arabic social media. Comput Speech Lang. 2014;28:20–37.

Web Search Engine-Based Representation for Arabic Tweets Categorization

Mohammed Bekkali and Abdelmonaime Lachkar

1 Introduction

Recently and with the explosion of use of social networks with over 2 billion active users all over the world, these social networks are an ideal collaborative tool for users looking for exchange, share, communicate and also have fun in a single place like Twitter, Facebook, etc. [1]. These social media allow users to post short texts. Twitter limits the length of each tweet to 140 characters. Facebook status length was limited to 420 characters before being extended. A personal status message on Windows Live Messenger is restricted to 128 characters. Most categories in Yahoo! Answers have an average post length less than 500 characters [2].

Short texts are different from traditional documents in its shortness and sparseness [3]. As a result, they tend to be ambiguous without enough contextual information. Thus the traditional text representation techniques are inadequate and may impact negatively in the accuracy of any Text Mining (TM) application such as clustering and categorization. To deal with the shortness and sparseness, most solutions proposed to enrich short texts representation aim to bring an additional semantics. The additional semantics could be from the short texts data collection itself or be derived from a much larger external knowledge base like Wikipedia, WordNet [4–6]. The former requires shallow Natural Language Processing (NLP) techniques while the later requires a much larger and appropriate datasets [3].

In this work, we propose an efficient method to enrich Arabic tweets representation based on web search engines which have become the most helpful tool for obtaining useful information from the web as a large and open corpus and also based

M. Bekkali (✉) • A. Lachkar
Department of Electrical and Computer Engineering, ENSA, Sidi Mohammed Ben Abbdellah University USMBA, Fez, Morocco
e-mail: mohammed.bekkali2@usmba.ac.ma; abdelmonaime.lachkar@usmba.ac.ma

© Springer International Publishing AG 2017

M. Kaya et al. (eds.), *From Social Data Mining and Analysis to Prediction and Community Detection*, Lecture Notes in Social Networks, DOI 10.1007/978-3-319-51367-6_4

on the Rough Set Theory (RST) [7, 8] which is a mathematical tool to deal with vagueness and uncertainty. On the one hand, by using the large volume of documents on the web, we can add more contextual terms that help to provide a greater context to the original tweets based on the snippets returned by the search engines. On the other hand, to select the terms that the original tweet will be enriched by from the returned snippets, we propose to use the RST which has been introduced by Pawlak in the early 1980s [7] and has been integrated in many TM techniques such as features selection. In this theory, each set in a Universe of objects is described by a pair of ordinary sets called Lower and Upper Approximations, determined by an equivalence relation in the Universe [7]. The RST will be presented with more details in Sect. 3.

In order to test and evaluate the effectiveness of our proposed method to enrich the Arabic tweets representation, we propose to develop an Arabic tweets categorization system. The reason behind this choice is that tweets are presented to the user in a chronological order [9]. This format of presentation is useful to the user since the latest tweets are generally more interesting than tweets about an event that occurred long time back. Merely, presenting tweets in a chronological order may be too embarrassing to the user, especially if he has many followers [9, 10]. Therefore, there is a great need to separate the tweets into different categories and then present the categories to the user. Text Categorization (TC) is a good way to overcome this problem.

TC systems try to find a relation between a set of texts and a set of categories. Machine Learning is the tool that allows deciding whether a text belongs to a set of predefined categories [10]. In the process of TC the document must pass through a series of steps (Fig. 1): transform the different types of documents into brut text, remove the stop words which are considered irrelevant words (prepositions and particles), and finally all words must be stemmed. Stemming is the process that consists of extracting the root from the word by removing the affixes [11–17]. To represent the internal of each document, the document must passed by the indexing process after preprocessing. Indexing process consists of three phases (Sebastiani [18]):

- All the terms appear in the documents corpus have been stocked in the super vector.
- Term selection is a kind of dimensionality reduction, it aims at proposing a new set of terms in the super vector to some criteria [19–21].
- Term weighting in which, for each term selected in phase (b) and for every document, a weight is calculated generally by TF-IDF which combines the definitions of term frequency and inverse document frequency [22, 23].

Finally, the classifier is built by learning the characteristics of each category from a training set of documents. After building of classifier, its effectiveness is tested by applying it to the test set and verifies the degree of correspondence between the obtained results and those encoded in the corpus.

In our work, we believe that Arabic tweets representation is a challenge and a crucial stage. It may impact positively or negatively on the accuracy of any tweets

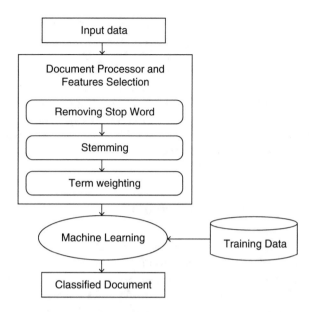

Fig. 1 Architecture of text categorization system

categorization system, and therefore the improvement of the representation step will lead by necessity to the improvement of Arabic tweets categorization system greatly.

The remainder parts of this paper are organized as follows: we begin with a brief survey on related work in enriching Arabic short texts representation and categorization in the next section. Section 3 describes some characteristics of the Arabic language and the stemming techniques. Section 4 presents the RST and its Tolerance Model to deal with textual information. Section 5 introduces the various Machine Learning techniques that have been exploited in this study. Section 6 describes our proposed system for enriching Arabic short texts representation. Section 7 conducts the experiments results. Finally, Sect. 8 concludes this paper and presents future work and some perspectives.

2 Related Work

2.1 Short Text Representation

Text representation enrichment in microblogs has become a crucial stage in the process of any text mining task such as Categorization, Opinion Mining, and Sentiments Analysis because of its shortness and sparseness. As a result, it has been intensively studied. Recently, Yunlun et al. propose a novel model for enriching the content of microblogs; first they build an optimization model to infer the topics of microblogs by employing the topic-word distribution of the external knowledge. Then the content of microblogs is further enriched by relevant words from external knowledge [24]. Jiliang et al. propose an efficient approach that enriches data representation by employing machine translation to increase the number of features from different languages. Then they propose a novel framework which performs multi-language knowledge integration and feature reduction simultaneously through matrix factorization techniques [3]. Hu et al. present a framework to improve the performance of short text clustering by mining informative context with the integration of Wikipedia and WordNet [5]. Finally, Banerjee et al. use the title and the description of news article as two separate query strings to select related concepts as additional feature [25].

2.2 Tweets Categorization

Recently, a number of papers have addressed the problem of tweets categorization most of them were tested against English tweets [9, 26]. Furthermore categorization systems that address Arabic tweets are very rare in the literature. Recently, Rehab et al. have recently presented a roadmap for understanding Arabic tweets through two main objectives [27]. The first is to predict tweet popularity in the Arab world. The second one is to analyze the use of Arabic proverbs in tweets; The Arabic proverbs classification model was labeled "Category" with four class values: sport, religious, political, and ideational.

3 Arabic Text Preprocessing

The Arabic language is the language of the Holy Quran. It is one of the six official languages of the United Nations and the mother tongue of approximately 300 million people. It is a Semitic language with 28 alphabet letters. Its writing orientation is from right-to-left. It can be classified into three types: Classical Arabic (العربية الفصحى), Modern Standard Arabic (العربية الحديثة), and Colloquial Arabic dialects (العربية العامية) [28].

Modern Standard Arabic has a rich morphology, based on consonantal roots, which depends on vowel changes and in some cases consonantal insertions and deletions to create inflections and derivations which make morphological analysis a very complex task [29]. There is no capitalization in Arabic, which makes it hard to identify proper names, acronyms, and abbreviations.

Stemming algorithms can be employed in Arabic Text Preprocessing to reduce words to their stems or roots. Very little research has been carried out on Arabic text. The nature of Arabic text is different than English text, and preprocessing of Arabic text is a more challenging stage in any application of Text Mining [23].

Arabic word stemming is a technique that aims to find the lexical root or stem for words in natural language, by removing affixes attached to its root, because an

Arabic word can have a more complicated form with those affixes. In Arabic there are three main approaches for stemming: Root-Based, Stem-Based approach, and Statistical ones.

Root-Based approach uses morphological analysis to extract the root of a given Arabic word. Many algorithms have been developed for this approach. Al-Fedaghi and Al-Anzi algorithm tries to find the root of the word by matching the word with all possible patterns with all possible affixes attached to it [11]. The algorithm does not remove any prefixes or suffixes. Al-Shalabi morphology system uses different algorithms to find the roots and patterns [12]. This algorithm removes the longest possible prefix, and then extracts the root by checking the first five letters of the word. This algorithm is based on an assumption that the root must appear in the first five letters of the word. Khoja has developed an algorithm that removes prefixes and suffixes, all the time checking that it's not removing part of the root and then matches the remaining word against the patterns of the same length to extract the root [30].

The aim of the Stem-Based approach is not to produce the root of a given Arabic word, rather is to remove the most frequent suffixes and prefixes. Light stemmer is mentioned by some authors affixes [14–17].

In statistical stemmer, related words are grouped based on various string similarities measures. Such approaches often involve n-gram [31]. Equivalence classes can be formed from words that share some initial letter n-gram or by refining these classes with clustering techniques. An n-gram is a set of n consecutive characters extracted from a word. The main idea behind this approach is that similar words will have a high proportion of n-grams in common. Typical values for n are 2 or 3 [23].

In this paper, we propose efficient method to enrich Arabic tweets representation by using web search engines combined with the RST. The RST will be presented in the following section.

4 Rough Set Theory

In this section we present the Rough Set Theory (RST) as a mathematical tool for imprecise and vague data, and his Tolerance Model to deal with Textual Information.

4.1 Generalized Approximation Spaces

Rough Set Theory has been originally developed as a tool for data analysis and classification [7, 8]. It has been successfully applied in various tasks, such as features selection/extraction, rule synthesis, and classification. The central point of Rough Set Theory is the notion of set approximation: any set in U (a nonempty set of object called the Universe) can be approximated by its Lower and Upper Approximation. In order to define Lower and Upper Approximation we need to introduce an indiscernibility relation that could be any equivalence relation R (reflexive, symmetric, and transitive). For two objects x, $y \in U$, if xRy, then we say that x and y are indiscernible from each other. The indiscernibility relation R induces a complete partition of universe U into equivalent classes $[x]_R$, $x \in U$ [32].

$$L_R(X) = \{x \in U : [x]_R \subseteq X\} \tag{1}$$

$$U_R(X) = \{x \in U : [x]_R \cap X \neq \Phi\} \tag{2}$$

Approximations can also be defined by means of rough membership function. Given rough membership function $\mu_X: U \rightarrow [0, 1]$ of a set $X \subseteq U$, the Rough Approximation is defined as:

$$L_R(X) = \{x \in U : \mu_X(x, X) = 1\} \tag{3}$$

$$U_R(X) = \{x \in U : \mu_X(x, X) > 0\} \tag{4}$$

Note that, given rough membership function as:

$$\mu X(x, X) = \frac{|[x]_R \cap X|}{|[x]_R|} \tag{5}$$

Rough Set Theory is dedicated to any data type but when it comes with documents representation we use its Tolerance Model described in the next section.

4.2 Tolerance Rough Set Model

Let $D = \{d_1, d_2 \ldots, d_n\}$ be a set of document and $T = \{t_1, t_2, \ldots, t_m\}$ set of index terms for D. With the adoption of the vector space model, each document d_i is represented by a weight vector $\{w_{i1}, w_{i2}, \ldots, w_{im}\}$ where w_{ij} denotes the weight of index term j in document i. The tolerance space is defined over a Universe of all index terms $U = T = \{t_1, t_2, \ldots, t_m\}$ [33].

Let $f_{di}(t_i)$ denotes the number of index terms t_i in document d_i; $f_D(t_i, t_j)$ denotes the number of documents in D in which both index terms t_i and t_j occur. The uncertainty function I with regard to threshold θ is defined as:

$$I_\theta = \{t_j | f_D(t_i, t_j) \geq \theta\} \cup \{t_i\} \tag{6}$$

Clearly, the above function satisfies conditions of being reflexive and symmetric. So $I_\theta(I_i)$ is the tolerance class of index term t_i. Thus we can define the membership function μ for $I_i \in T$, $X \subseteq T$ as (Lang 2003):

$$\mu_X (t_i, X) = v (I_\theta (t_i), X) = \frac{|I_\theta (t_i) \cap X|}{|I_\theta (t_i)|} \tag{7}$$

Finally, the Lower and the Upper Approximation of any document $d_i \subseteq T$ can be determined as:

$$L_R (d_i) = \{t_i \in T : v (I_\theta (t_i), d_i) = 1\} \tag{8}$$

$$U_R (d_i) = \{t_i \in T : v (I_\theta (t_i), d_i) > 0\} \tag{9}$$

In the following section, we will present our proposed method to enrich Arabic tweet's representation using the web as a large and open corpus combined with the Rough Set Theory.

5 Machine Learning for Text Categorization

TC is the task of automatically sorting a set of documents into categories from a predefined set. This section covers three algorithms among the most used Machine Learning Algorithms for TC: Naïve Bayesian (NB), Support Vector Machine (SVM), and Decision Tree (DT).

5.1 Naïve Bayesian

The NB is a simple probabilistic classifier based on applying Baye's theorem, and it's powerful, easy, and language independent method [34]. When the NB classifier is applied on the TC problem we use Eq. (10):

$$p (\text{class}|\text{doc}) = \frac{p (\text{class}) . p (\text{doc}|\text{class})}{p (\text{doc})} \tag{10}$$

where:

- $p(\text{class}|\text{doc})$: It's the probability that a given document D belongs to a given class C
- $p(\text{doc})$: The probability of a document, it's a constant that can be ignored
- $p(\text{class})$: The probability of a class, it's calculated from the number of documents in the category divided by documents number in all categories
- $p(\text{doc}|\text{class})$: it's the probability of document given class, and documents can be represented by a set of words:

$$p (\text{doc}|\text{class}) = \prod_i p (\text{word}_i|\text{class}) \tag{11}$$

so:

$$p\,(\text{class}|\text{doc}) = p\,(\text{class})\,.\prod_i p\,(\text{word}_i|\text{class}) \tag{12}$$

where:

$p(\text{word}_i|\text{class})$: The probability that a given word occurs in all documents of class C, and this can be computed as follows:

$$p\,(\text{word}_i|\text{class}) = \frac{T_{\text{ct}} + \lambda}{N_c + V} \tag{13}$$

where:

- T_{ct}: The number of times that the word occurs in that category C.
- N_c: The number of words in category C.
- V: The size of the vocabulary table.
- λ: The positive constant, usually 1 or 0.5 to avoid zero probability.

5.2 Support Vector Machine

SVM introduced by [35] and has been introduced in TC by [36]. Based on the structural risk minimization principle from the computational learning theory, SVM seeks a decision surface to separate the training data points into two classes and makes decisions based on the support vectors that are selected as the only effective elements in the training set.

Given a set of N linearly separable points $N = \{x_i \in R_n | i = 1, 2, \dots, N\}$, each point x_i belongs to one of the two classes, labeled as $y_i \in \{-1, 1\}$. A separating hyper-plane divides S into two sides, each side containing points with the same class label only. The separating hyper-plane can be identified by the pair (w, b) that satisfies: $w\cdot x + b = 0$

and:

$$\begin{cases} w \cdot x_i + b \geq +1 \text{ if } y_i = 1 \\ w \cdot x_i + b \leq -1 \text{ if } y_i = -1 \end{cases} \tag{14}$$

For $i = 1, 2, \dots, N$; where the dot product operation $(\cdot)$ is defined by:

$$w \cdot x = \sum w_i \cdot x_i$$

For vectors w and x, thus the goal of the SVM learning is to find the Optimal Separating Hyper (OSH) plane that has the maximal margin to both sides. This can be formularized as:

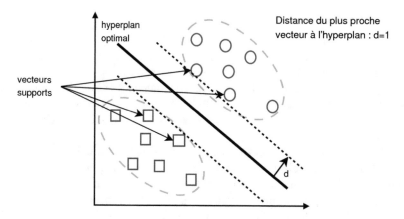

Fig. 2 The optimal separating hyper-plane

minimize: ½ ·w·w
subject to:

$$\begin{cases} w \cdot x_i + b \geq +1 \text{ if } y_i = 1 \\ w \cdot x_i + b \leq -1 \text{ if } y_i = -1 \end{cases} \qquad (15)$$

Figure 2 shows the optimal separating hyper-plane.

The small crosses and circles represent positive and negative training examples, respectively, whereas lines represent decision surfaces. Decision surface σ_i (indicated by the thicker line) is, among those shown, the best possible one, as it is the middle element of the widest set of parallel decision surfaces (i.e., its minimum distance to any training example is maximum). Small boxes indicate the Support Vectors.

During classification, SVM makes decision based on the OSH instead of the whole training set. It simply finds out on which side of the OSH the test pattern is located. This property makes SVM highly competitive, compared with other traditional pattern recognition methods, in terms of computational efficiency and predictive accuracy [37].

The method described is applicable also to the case in which the positives and the negatives are not linearly separable. Yang and Liu experimentally compared the linear case (namely, when the assumption is made that the categories are linearly separable) with the nonlinear case on a standard benchmark, and obtained slightly better results in the former case [37].

5.3 *Decision Tree*

Decision tree is a classification technique. It is a tree like structure where internal node contains splits and splitting attributes. It represents test on an attribute. Arcs between internal node and its child contain consequences of test. Each leaf node is associated with a class label. Decision tree is constructed from training set. Then this decision tree is used to classify the tuples with unknown class label [38, 39] (Fig. 3).

The goal is create a model to predict value of target variable based on input values. Training dataset is used to create tree and test dataset is used to test accuracy of the decision tree. Each leaf node represents the target attribute's value depend on input variables represented by path from root to leaf node. First, an attribute that splits data efficiently is selected as root node in order to create small tree. The attribute with higher information is selected as splitting attribute [40].

Decision tree algorithm involves three steps:

1. For a given dataset *S*, select an attribute as target class to split tuples into partitions.
2. Determine a splitting criterion to generate a partition in which all tuples belong to a single class. Choose best split to create a node.
3. Iteratively repeat above steps until complete tree is grown or any stopping criterion is fulfilled (Fig. 4).

In the following section we present our proposed method for enriching Arabic short texts.

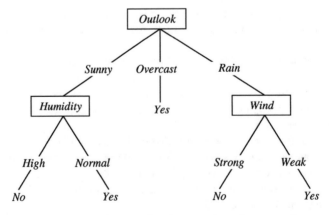

Fig. 3 Decision tree showing whether to go for trip or not depending on weather

Fig. 4 Decision tree induction

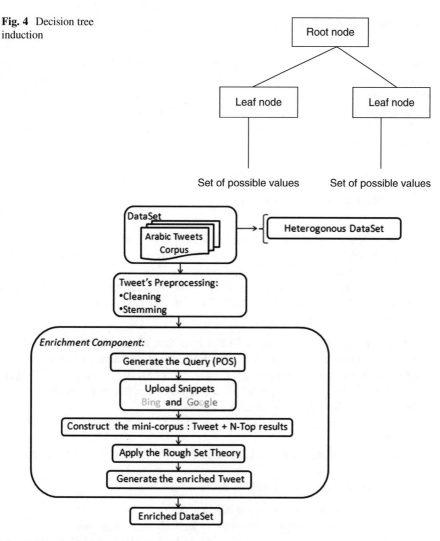

Fig. 5 Flowchart of our proposed method for Arabic tweets representation enrichment

6 Proposed Method for Enriching Arabic Tweets Representation

In this section, we present in detail our proposed method for enriching Arabic tweets representation. Figure 5 presents a flowchart showing the steps of our proposed system for Arabic tweets representation enrichment.

6.1 Tweet's Preprocessing

Each tweet in the corpus will be cleaned by removing Arabic stop words, Latin words, and special characters like (/, #, $, etc.). After that we apply a stemmer algorithm to generate the root/stem and eliminate the redundancy. Different algorithms have been proposed in this topic such as Khoja [30] and Larkey Stemmer [17] which both are the most known algorithms for Arabic text preprocessing. We used Khoja Stemmer in our system.

6.2 Enrichment Component

This component presents the process of enriching the Arabic tweet's representation; in our contribution we enrich the tweet's representation by additional terms existing in the n-top retrieved results returned by two of the most popular web search engines, Google and Bing, by sending the content of the tweet as a query to the web search engines. Each tweet in the DataSet will pass through the following treatment:

- Most of tweets contain between 10 and 30 terms; if we send the entire content of the tweet as a query, the web search engine may return an empty list. To overcome this problem, we propose to use a Part of Speech Tagging (POSTagger) to reduce the length of the query by sending only the nouns of the tweet; we used the Stanford POSTagger [41] to this purpose
- Upload the n-top retrieved results without redundancy based on the URL of each snippet
- Construct a new and enriched mini-corpus formed by the original tweet and the n-top retrieved snippets
- Applying the Rough Set Theory by the following steps:
  - Generate the terms set from the mini-corpus
  - Preprocess all the snippets returned by the web search engines
  - Calculate the frequency of each term in the tweet and in the whole mini-corpus
  - Determine the tolerance class of all terms; for a given term, the tolerance class contains all the terms that occur with this term in the same tweet a number of times upper than θ. This tolerance class is defined using the formula (6)
  - Deduce the Upper Approximation of the tweet by using the formula (9) then for each term a weight is calculated by replacing the original TF-IDF formula which combines the definitions of term frequency and inverse document frequency by the following formula:

$$w_{ij} = \begin{cases} (1 + \ln(f_{di}(t_i))) * \ln \frac{N}{f_D(t_j)} & \cdots \; ; t_j \in d_i \\[2em] \min t_k \in w_{ij} * \dfrac{\ln\left(\frac{N}{f_D(t_j)}\right)}{1+\ln\left(\frac{N}{f_D(t_j)}\right)} & \cdots \; ; t_j \in U_R \frac{(d_i)}{d_i} \\[2em] 0 & ; t_j \notin U_R(d_i) \end{cases} \tag{16}$$

where w_{ij} is the weight for term j in the document d_i. This formula ensures that each term occurring in the Upper Approximation of d_i but not in d_i has a weight smaller than the weight of any terms in d_i. Normalization by vector's length is applied to all weights of document vectors w_{ij} [32].

$$w_{ij} = \frac{w_{ij}}{\sqrt{\sum t_K \in U_R(d_i)(w_{ij})^2}} \tag{17}$$

After we take back the same process for all the tweets in the initial corpus, we have a new enriched corpus ready to be classified. The enrichment process algorithm can be summarized and presented as below (Table 1).

Finally, in order to test and evaluate the effectiveness of our proposed method to enrich the Arabic tweets representation, we propose to develop an Arabic tweets categorization system. The classifier is built by learning the characteristics of each category from a training set of tweets. After building of classifier, its effectiveness is tested by applying it to the test set and verifies the degree of correspondence between the obtained results and those encoded in the corpus. In our test we have been used NB, SVM, and DT as classifiers.

7 Experiments Results

In this work, to illustrate that our proposed method can improve the Arabic tweet's representation by using the web search engines and the upper approximation of the RST and therefore can enhance the performance of any Arabic tweets categorization system. In addition, to contrast the effect of using the web as a large and an open corpus, we have been enriched the Arabic tweets based only on the RST by adding to the original tweets terms that are semantically related in the same static corpus. In this section a series of experiments has been conducted using two different and diverse datasets so as to test and evaluate the interest of our contribution.

Figure 6 describes our experiments. Firstly, we categorize the originals tweets and after that we have been generated two enriched dataset, the first using only the RST and the second one based on the web search engine and RST.

Table 1 Enrichment process algorithm

```
Begin
    Initialise the tweet's Data Set
    for each tweet twᵢ the Data Set
        query = nounsOf(twᵢ)
        List<Snippets>  res = NTopResults(query)
        Corpus u = twᵢ U res
        generateUpperApproximation(twᵢ, u)
        addUpperApproximationToEnrichedDataSet(ds, twᵢ)
    end for
end
void generateUpperApproximation(tw, u)
    Preprocessing(u)
    Initialise the TermSet s
    Initialise the co-occurrence matrix
    for each Term tᵢ in s do
        for each Term tⱼ in s do
            if occurTogether(tᵢ, tⱼ) > θ then
                addToToleranceClass(tᵢ, tⱼ)
        end for
    end for
    for each Term tⱼ in tw do
        tct = toleranceClass(tⱼ)
        cof = |tct ∩ twᵢ| / |tct|
        if cof > 0 then
            addTermToUpperApprox(tw, tⱼ)
    end for
end
```

7.1 Datasets

Tweets in our experiments are collected from Twitter by the NodeXL Excel Template which is a freely Excel template that makes it super easier to collect Twitter network data [42]. The two datasets are manually classified into eight categories (Table 2). These categories are:

- Dataset_1: ("Cinema"/"السينما"), ("Documentary"/"وثائقي"), ("Economics"/ "الاقتصاد"), ("Education"/"التعليم"), ("Health"/"الصحة"), ("News"/"الأخبار"), ("Sport"/"الرياضة"), and ("Tourism"/"السياحة")
- Dataset_2: ("Economics"/"الاقتصاد"), ("Electronics"/"الإلكترونيك"), ("Health"/"الصحة"), ("Politic"/"السياسة"), ("Religion"/"الدين"), ("Science"/ "العلوم"), ("Sociology"/"علم الإجتماع"), and ("Sport"/"الرياضة").

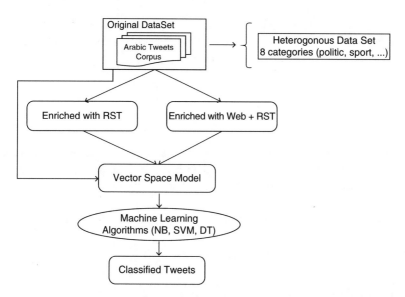

Fig. 6 Descriptions of our experiment's description

Table 2 Dataset's
description

	Categories	Number of tweets
Dataset_1	Cinema	240
	Documentary	188
	Education	147
	Economy	131
	Health	320
	News	120
	Sport	139
	Tourism	196
	Total	1481
Dataset_2	Economy	682
	Electronics	160
	Health	580
	Politic	154
	Religion	628
	Science	540
	Sociology	223
	Sport	508
	Total	3475

In contrast, for uploading the snippets the query will be sent to both Google and Bing Web Search Engines using the services offered by the Google API [43] and Bing API [44].

Each dataset has been divided into two parts: training and testing. The training data consist of 75% of the tweets per category. The testing data, on the other hand, consist of 25% of the tweets for each category.

7.2 Results

To assess the performance of the proposed system, a series of experiments has been conducted. The effectiveness of our system has been evaluated and compared in terms of the F1-measure using the NB, SVM, and DT classifiers in our Arabic tweets categorization system.

F1-measure can be calculated using Precision and Recall measures as follows:

$$F1 - measure = 2 * \frac{(P * R)}{(P + R)} \tag{18}$$

Precision and Recall both are defined, respectively, as follows:

$$P = \frac{TP}{(TP + FP)} \tag{19}$$

$$R = \frac{TP}{(TP + FN)} \tag{20}$$

where:

- True Positive (TP) refers to the set of tweets which are correctly assigned to the given category.
- False Positive (FP) refers to the set of tweets which are incorrectly assigned to the category.
- False Negative (FN) refers to the set of tweets which are incorrectly not assigned to the category.

On the other hand, Table 3 presents some examples to illustrate the semantic links discovered by the RST. The first one presents a tweet about economy that contains the term ("oil"/"النفط"). The term ("minister"/"وزير") was added to the generated Upper Approximation which comes together in some contexts. The second example presents a tweet about education that contains the term ("ministry"/"وزارة"), the two terms ("higher"/"العالي") and ("education"/"التعليم") were added to the generated Upper Approximation which refers to the ministry of higher education.

The obtained results for Precision, Recall, and F1-measure concerning the Dataset_1 (respectively, Dataset_2) are presented in Figs. 7, 8, and 9 (respectively, Figs. 10, 11, and 12).

Table 3 Examples of tweets after enrichment

Category	Tweet's content		Initial terms	Added terms
Economy/ الاقتصاد	Arabic	أعلى إلى الذهب صعود أسابيع ثلاثة مستوياته موافقة بعد ونصف توجيه على واشنطن العراق على جوية	أعلى، الذهب، صعود، أسابيع، ثلاثة، مستويات، واشنطن، موافقة، نصف، العراق جوية، ضربات،	عالم، وزير
	English translation	The rise of gold to its highest level, three and a half weeks after the approval of Washington to direct flights on Iraq	Rise, gold, top, levels, three, weeks, half, approval, Washington, strikes, air, Iraq	World, minister
Education/ التعليم	Arabic	تنفذالوزارة لذوي منسوبيهابرنامج الخاصةالاحتياجات	مج، برنا تنفذ، وزارة، ذوي، منسوبيها، خاص الاحتياجات،	تعليم، العالي
	English translation	The ministry implement a program for employees with special needs	Ministry, implement, program, employees, needs, special	Education, higher

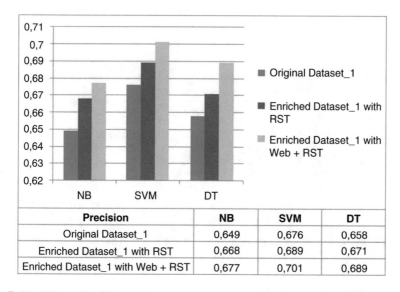

Precision	NB	SVM	DT
Original Dataset_1	0,649	0,676	0,658
Enriched Dataset_1 with RST	0,668	0,689	0,671
Enriched Dataset_1 with Web + RST	0,677	0,701	0,689

Fig. 7 Precision results of dataset_1

7.3 Discussion

For example, when using NB classifier and the second dataset (Dataset_2): the Precision with the original tweets is 0.776, in contrast when we enrich their content, the obtained Precision is 0.785 (using RST) and 0.834 (using web and RST); we remark that improving the tweets representation using both the web and RST impact

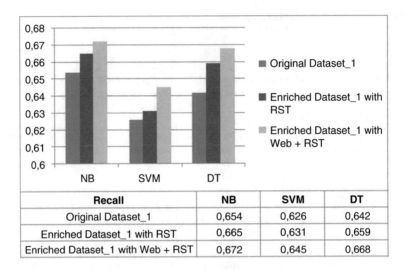

Fig. 8 Recall results of dataset_1

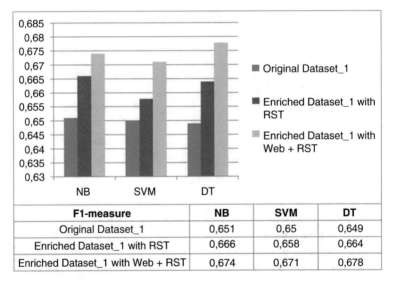

Fig. 9 F1-measure results of dataset_1

positively on the Precision of our Arabic tweets categorization system more than using just the RST on our static corpus. These results demonstrate that the returned snippets contain more relevant terms that have a positive influence on the process of categorization.

The Recall with the original tweets is 0.767, in contrast when we enrich their content, the obtained Recall is 0.795 (using RST) and 0.817 (using web and

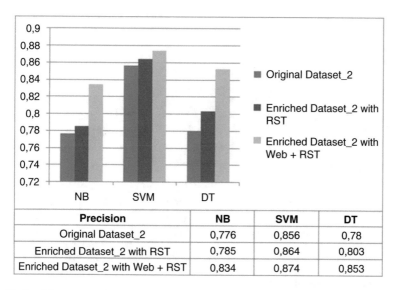

Fig. 10 Precision results of dataset_2

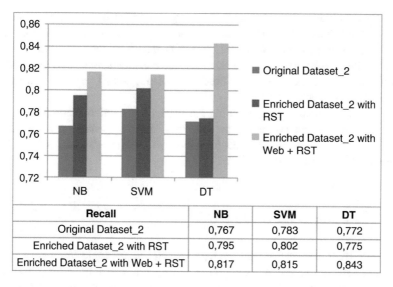

Fig. 11 Recall results of dataset_2

RST); these results illustrate that adding more contextual terms help to provide a greater context to the original tweets and therefore decrease the effect of sparseness problem.

The F1-measure with the original tweets is 0.771, in contrast when we enrich their content, the obtained Recall is 0.789 (using RST) and 0.825 (using web and

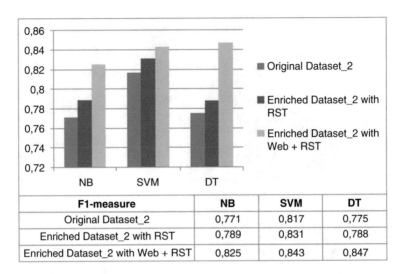

F1-measure	NB	SVM	DT
Original Dataset_2	0,771	0,817	0,775
Enriched Dataset_2 with RST	0,789	0,831	0,788
Enriched Dataset_2 with Web + RST	0,825	0,843	0,847

Fig. 12 F1-measure results of dataset_2

RST); we conclude that enriching the tweets representation by additional terms from the web can enhance the performance of our Arabic tweets categorization system more than enriching their content only by terms existing in the same corpus.

In addition, our proposed method has been evaluated using SVM and DT classifiers. The obtained results for Precision, Recall, and F1-measure demonstrate the interest of our contribution.

Finally, we conclude that enriching the tweets representation improves greatly our Arabic tweets categorization system, it is either by using only the RST or combined with the returned snippets from the web search engines so as to provide a greater context to the original tweets.

8   Conclusion and Future Work

Improving short texts representation becomes an interest topic in recent years. In order to overcome its shortness and sparseness, various solutions have been tried to enrich short text to get more features, including using the web search engines or external knowledge like Arabic WordNet (AWN) or Wikipedia. These methods solve the problem to certain extent, but still leave much space for improvement.

In this paper, we have proposed an effective method for enriching Arabic tweets representation based on web search engines and the RST. This latter enriches and adds other terms which are semantically related with the original terms existing in the original tweets. The proposed method has been integrated and tested for an Arabic tweets categorization using NB, SVM, and DT classifiers. The obtained results demonstrate the interest of our contribution.

In our future work, we will focus on using a tweet's representation based on concepts instead of terms.

References

1. Kemp S. Global-social-media-users-pass-2-billion. 2015. http://wearesocial.net/blog/2014/08/global-social-media-users-pass-2-billion/. Accessed Dec 2015
2. Adamic LA, Zhang J, Bakshy E, Ackerman MS. Knowledge sharing and yahoo answers: everyone knows something. In: Proceedings of 17th International Conference on World Wide Web; 2008. New York: ACM. pp. 665–74
3. Jiliang T, Xufei W, Huiji G, Xia H, Huan L. Enriching short text representation in microblog for clustering front. Comput Sci. 2012;6(1) doi:10.1007/s11704-009-0000-0.
4. Phan XH, Nguyen LM, Horiguchi S. Learning to classify short and sparse text & web with hidden topics from large-scale data collections. Proceedings of the 17th International Conference on World Wide Web; 2008. New York: ACM. pp. 91–100
5. Hu X, Sun N, Zhang C, Chua TS. Exploiting internal and external semantics for the clustering of short texts using world knowledge. Proceedings of the 18th ACM Conference on Information and Knowledge Management; 2009. New York: ACM. pp. 919–28
6. Chen M, Jin X, Shen D. Short text classification improved by learning multigranularity topics. Proceedings of the 22nd International Joint Conference on Artificial Intelligence; 2011. Barcelona: Citeseer. pp. 1776–81
7. Pawlak Z. Rough sets: theoretical aspects of reasoning about data. Dordrecht: Kluwer; 1991.
8. Komorowski J, Polkowski L, Skowron A. Rough sets: A tutorial. Singapore: Springer-Verlag; 1998.
9. Sriram B, Fuhry D, Demir E, Ferhatosmanoglu H. Short Text Classification in Twitter to Improve Information Filtering, SIGIR'10, 19–23 July 2010; Geneva, Switzerland. ACM 978-1 60558-896-4/10/07
10. Sebastiani F. Machine learning in automated text categorization. ACM Comput Surv. 2002;34(1):1–47.
11. Al-Fedaghi S, Al-Anzi F. A new algorithm to generate Arabic root-pattern forms. In: Proceedings of the 11th National Computer Conference and Exhibition; 1989. pp. 391–400
12. Al-Shalabi R, Evens M. A computational morphology system for Arabic. In: Workshop on Computational Approaches to Semitic Languages, COLING-ACL98; 1998
13. Khoja S. Stemming arabic text. Lancaster: Computing Department, Lancaster University; 1999.
14. Larkey L, Connell ME. Arabic information retrieval at UMass in TREC-10. Proceedings of TREC 2001, Gaithersburg: NIST; 2001
15. Aljlayl M, Frieder O. On Arabic search: improving the retrieval effectiveness via a light stemming approach. Proceedings of ACM CIKM 2002 International Conference on Information and Knowledge Management. McLean, VA: ACM; 2002. pp. 340–7
16. Chen A, Gey F. Building an Arabic stemmer for information retrieval. In Proceedings of the 11th Text Retrieval Conference (TREC 2002), National Institute of Standards and Technology; 2002
17. Larkey L., Ballesteros L, Connell ME, Improving stemming for Arabic information retrieval: light stemming and co-occurrence analysis. Proceedings of SIGIR'02; 2002. New York: ACM. pp. 275–82
18. Sebastiani F. A tutorial on automated text categorisation. Proceedings of ASAI-99, 1st Argentinian Symposium on Artificial Intelligence; 1999. Buenos Aires: Citeseer. pp. 7–35

19. Yang Y, Pedersen JO. A comparative study on feature selection in text categorization. Proceedings of ICML-97. 1997. San Francisco: Morgan Kaufmann Publishers Inc. pp. 412–20
20. Rogati M, Yang Y. High-performing feature selection for text classification. CIKM'02, ACM; 2002
21. Liu T, Liu S, Chen Z, Ma WY. An evaluation on feature selection for text clustering. Proceedings of the 12th International Conference (ICML 2003). Washington, DC; 2003. pp. 488–95
22. Aas K, Eikvil L. Text categorisation: a survey. Technical report, Norwegian Computing Center; 1999
23. Hadni M, Lachkar A, Alaoui OS. Effective Arabic stemmer based hybrid approach for Arabic text categorization. Int J Data Min Knowl Manag Process (IJDKP). 2013;3(4):1.
24. Yang Y, Deng Z, Yu H. A novel content enriching model for microblog using news corpus. Proceedings of the 52nd Annual Meeting of the Association for Computational Linguistics (Short Papers); 2014. Baltimore: ACM. pp. 218–23
25. Banerjee S, Ramanathan K, Gupta A. Clustering short texts using Wikipedia. Proceedings 30th annual international ACM SIGIR conference on Research and development in information retrieval; 2007. New York: ACM. pp. 787–8
26. Antenucci D, Handy G, Modi A, Tinkerhess M. Classification of tweets via clustering of hashtags. EECS 545 FINAL PROJECT, FALL; 2011
27. Nasser Al-Wehaibi R, Khan MB. Understanding the content of Arabic tweets by data and text mining techniques. Symposium on Data Mining and Applications; 2014
28. Froud H, Lachkar A, Ouatik SA. A comparative study of root-based and stem-based approaches for measuring the similarity between Arabic words for Arabic text mining applications. Adv Comput Int J (ACIJ). 2012;3(6):55.
29. Abu-Hamdiyyah M. The Qur'An: An introduction. London: Routledge; 2000.
30. Khoja S, Garside R. Stemming Arabic text. Lancaster: Computer Science Department, Lancaster University; 1999.
31. Khreisat L. Arabic text classification using N-gram frequency statistics a comparative study. Proceedings of the International Conference on Data Mining; 2006. Las Vegas: USCCM. pp. 78–82
32. Chi Lang N. A tolerance rough set approach to clustering web search results. Poland: Warsaw University; 2003.
33. Zhang J, Chen S. A study on clustering algorithm of Web search results based on rough set. Software Engineering and Service Science (ICSESS); 2013
34. Alsaleem S. Automated Arabic text categorization using SVM and NB. Int Arab J e-Technol. 2011;2(2):124.
35. Vapnik V. The nature of statistical learning theory, chapter 5. New York: Springer-Verlag; 1995.
36. Joachims T. Text categorization with support vector machines: learning with many relevant features. In: Proceedings of the European Conference on Machine Learning (ECML); 1998. Chemnitz: Springer-Verlag. pp. 137–42
37. Yang Y, Liu X. A re-examination of text categorization methods. Proceedings of the 22nd Annual International ACM SIGIR Conference on Research and Development in Information Retrieval (SIGIR'99), 1999. Berkeley: ACM. pp. 42–49
38. Kaur D, Bedi R, Gupta SK. Review of decision tree data mining algorithms: Id3 and C4.5. Proceedings of International Conference on Information Technology and Computer Science; 11–12 July 2015
39. Kabra RR, Bichkar RS. Performance prediction of engineering students using decision tree. Int J Comput Appl. 2011;36(11):8–12.
40. Kesavraj G, Sukumaran S. A study on classification technique in data mining. 4th ICCNT-2013; 2013
41. Toutanova K, Klein D, Manning C, Singer Y. Feature-rich part-of-speech tagging with a cyclic dependency network. In Proceedings of HLT-NAACL 2003. pp. 252–9

42. Lamberson PJ. Collecting and visualizing twitter network data with NodeXl and Gephi. http://social-dynamics.org/twitter-network-data/. Accessed Dec 2015
43. https://developers.google.com/custom-search/docs/start
44. https://datamarket.azure.com/dataset/5BA839F1-12CE-4CCE-BF57-A49D98D29A44

Sentiment Trends and Classifying Stocks Using P-Trees

Arijit Chatterjee and William Perrizo

1 Introduction

Twitter is one of the most important social media platforms in today's world providing the unique ability for a user to connect with almost anyone else in the world. The platform supports 33 languages and has close to 288 million active users and almost 500 million tweets are created every day. Investors are day by day using Twitter platform to cite their opinions on particular ticker symbols and share their market focused posts and updates. But with so much of information in the platform it is really difficult to find the information, a particular user needs to make an investment decision. In the realm of stocks it is important to understand which ticker symbols to follow, based on which investment decisions can be made.

The solution described in the paper will provide them with the ability to identify the ticker symbols effectively without reading through huge corpuses of tweet texts. Now once the ticker symbols with the most frequent occurrences from the tweets are identified, the paper gives an insight into the tweet texts associated with the ticker symbols. The paper also shows how investors can be biased and its effect on the behavior of the stocks in the market. Section 1 of the paper discusses the benefit of using Twitter as the social media platform, Sect. 2 of the paper introduces the concept of P-Trees and how we can structure our data vertically, Sect. 3 of the paper shows how we can store the necessary information in the data cube model and

A. Chatterjee (✉)
Department of Computer Science, North Dakota State University, Fargo, ND, USA

Microsoft Research and Development, One Microsoft Way, Redmond, WA, USA
e-mail: arijit.rony@hotmail.com

W. Perrizo
Department of Computer Science, North Dakota State University, Fargo, ND, USA
e-mail: william.perrizo@ndsu.edu

© Springer International Publishing AG 2017
M. Kaya et al. (eds.), *From Social Data Mining and Analysis to Prediction and Community Detection*, Lecture Notes in Social Networks,
DOI 10.1007/978-3-319-51367-6_5

represent various views from the captured information. Section 4 shows results on classifying ticker symbols from the tweets of multiple investors and an insight to remove the investor's bias based on tweet counts. Section 5 of the paper introduces sentiment analysis and the meaningful features for building classifiers while Sect. 6 introduces Microsoft's Azure Sentiment Analyzer and the comparison with the other NLP tools. Section 7 of the paper discusses the hypothesis around sentiment trends relevant to this research. The future direction of this work is shown in Sect. 8 and we finally conclude in Sect. 9.

2   The P-Tree Technology

Tremendous volumes of data cause the cardinality problem for conventional transaction based data mining algorithms. For fast and efficient data processing, we transform the data into P-Tree, the loss-less, compressed, and data-mining-ready vertical data structure.

The basic data structure exploited in the P-Tree technology [1] is the Predicate Count Tree[1] (PC-Tree) or simply the P-Tree. Formally, P-Trees are tree-like data structures that store numeric-relational data (i.e., numeric data in relational format) in column-wise, bit-compressed format by splitting each attribute into bits (i.e., representing each attribute value by its binary equivalent), grouping together all bits in each bit position for all tuples and representing each bit group by a P-Tree. P-Trees provide a lot of information and are structured to facilitate data mining processes. After representing each numeric attribute value by its bit representation, we store all bits for each position separately. In other words, we group together all the bit values at bit position x of each attribute for all tuples t. Figure 1 shows a relational table made up of three attributes and four tuples transformed from numeric to binary, and highlights all the bits in the first three bit groups for the first attribute, Attribute 1; each of those bit groups will form a P-Tree. Since each attribute value in our table is made up of 8 bits, 24 bit groups are generated in total with each attribute generating 8 bit groups. Figure 2 shows a group of 16 bits transformed into a P-Tree after being divided into quadrants (i.e., subgroups of 4). Each such tree is called a basic P-Tree. In the Fig. 2, 7 is the total number of 1's in the whole bit group shown in the upper part of Fig. 2. 4, 2, 1, and 0 are the number of 1's in the 1st, 2nd, 3rd, and 4th quadrants, respectively, in the bit group. Since the first quadrant (the node denoted by 4 on the second level in the tree) is made up of "1" bits in its entirety (we call it a pure-1 quadrant) no sub-trees for it are needed. Similarly, quadrants made up entirely of "0" bits (the node denoted by "0" on the second level in the tree) are called pure-0 quadrants and have no sub-trees. As a matter of

---

[1]Formerly known as the Peano Count Tree.

Attribute1	Attribute2	Attribute3	Attribute1	Attribute2	Attribute3
5	6	2	00000101	00000110	00000010
12	15	0	00001100	00001111	00000000
14	24	1	00001110	00001000	00000001
29	64	255	00011101	01000000	11111111

Fig. 1 Relational numeric data converted to binary format with the first three bit groups in attribute 1 highlighted

Fig. 2 A 16 bit group converted to a P-Tree

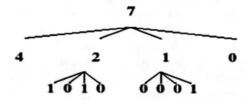

fact, this is how compression is achieved[2] [1]. Non-pure quadrants such as nodes 2 and 1 on the second level in the tree are recursively partitioned further into four quadrants with a node for each quadrant. We stop the recursive partitioning of a node when it becomes pure-1 or pure-0 (eventually we will reach a point where the node is composed of a single bit only and is pure because it is made up entirely of either only "1" bits or "0" bits). P-Tree algebra includes operations such as AND, OR, NOT (or complement), and ROOTCOUNT (a count of the number of 1's in the tree). Details for those operations can be found in [1]. The latest benchmark on P-Trees AND'ing has shown a speed of 6 ms for AND'ing two P-Trees representing bit groups each containing 16 million bits. Speed and compression aspects of P-Trees have been discussed in greater details in [1–4] that gives some applications exploiting the P-Tree technology. Once we have represented our data using P-Trees, no scans of the database are needed to perform text categorization. In fact, this is one of the important aspects of P-Tree technology.

---

[2]It's worth noting that good P-Tree compression can be achieved when the data is very sparse (which increases the chances of having long sequences of "0" bits) or very dense (which increases the chances of having long sequences of "1" bits).

3   Data Cube Model and Views

In this section we describe how we can represent the captured tweets in the form of P-Trees and store the data lossless in a 3-D data cube model [5, 6].

In the term space model [7, 8], a document is presented as a vector in the term space where terms are used as features or dimensions. The data structure resulting from representing all the documents in a given collection as term vectors is referred to as a document-by-term matrix. Given that the term space has thousands of dimensions, most current text-mining algorithms fail to scale-up. This very high dimensionality of the term space is an idiosyncrasy of text mining and must be addressed carefully in any text-mining application.

Within the term space model, many different representations exist. On one extreme, there is the binary representation in which a document is represented as a binary vector where a 1 bit in slot i implies the existence of the corresponding term ti in the document in position p_j, and a 0 bit implies its absence. This model [9, 10] is fast and efficient to implement but clearly lacks the degree of accuracy needed because most of the semantics are lost.

On the other extreme, there is the frequency representation where a document is represented as a frequency vector [7, 8]. Many types of frequency measures exist: *term frequency* (TF), *term frequency by inverse document frequency* (TFxIDF), *normalized TF*, and the like. This representation for term frequency is obviously more accurate than the binary one but is not as easy and efficient to implement. Our algorithm shows the *term frequency* (TF) measure based on ticker symbol references in the tweet texts and is characterized by accuracy and space and time efficiency because it is based on the P-Tree technology.

For our analysis we store the tweet ID, tweet Text, and User Information as different dimensions of every tweet. We also store time as a supplemental information which can be used later in analysis. In a data cube model, the cubes can be used to represent multidimensional data of different dimensions. Each cube cells denotes an event while the cube edges stand for analysis dimensions. Each cube cell is given a value of each measure. So far, we've created a binary matrix similar to that depicted in Fig. 1. We follow the same steps presented in Sect. 3 to create the P-Tree version of the matrix as in Fig. 3. For every bit position in every term ti we will create a basic P-Tree. $P_{i,j}$ is the P-Tree representation of the bits lying at jth position in ith term for all documents. Each $P_{i,j}$ will give the number of documents having a 1 bit in position j for term i. This representation conveys a lot of information and is structured to facilitate fast data mining processing. To get the P-Tree holding all documents having a certain value for some term i, we can follow the steps given in the following example: if the desired binary value for term i is 10, we calculate $P_{i,10}$ as $P_{i,10} = P_{i,1}$ AND $P'_{i,0}$ where "'" indicates the bit-complement or the NOT operation (which is simply the count complement in each quadrant).

For every tweet made by a user, we parse the tweet Text and store all the unique words in a list. At the same time, we also find the max Position of the words in the tweet. Twitter has a 140 character limit for every tweet, so the max Position of the

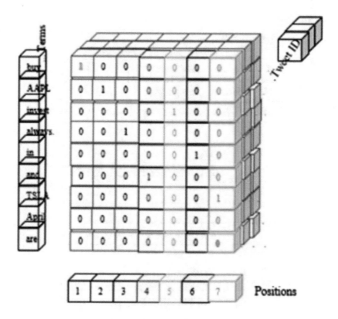

Fig. 3 Data cube diagram showing the dimension of unique term, position, and tweet IDs

words in most cases are found to be less than 35. Every user related dataset can be thought to be captured as an aggregate of tweet IDs indexed on the unique word occurrences from their tweets and the respective positions of these unique terms in the tweets.

The above cube shows the representation of a single tweet text "Buy AAPL always and invest in TSLA" where based on the unique terms and their positions we denote a 1 where there is an occurrence else a 0. The data for the ticker symbols with reference to $<stocksymbol> in tweets are all stored in the data cube and since all the bit slices in the P-Trees are 1 and 0 the total 1 count at the end of the bit slice generates the frequency of occurrence for the ticker symbol. Now based on this cube there can be different forms of P-Tree views which can be created based on how we would like to slice the data cube. AND or OR logical operations when performed on bit slices offer significant computational benefit when tuples are million records deep as this is the way in an assembly level a processor computes. This might not seem apparent when the dataset is small but for larger datasets P-Tree based logical operations provide significant improvements. Intermediate to the creation of P-Trees we view the data cube as tables three ways: the Document Table (DT), the Term Table (TT), and the Position Table (PT) as shown in Table 1.

Table 1 : Data cube tables

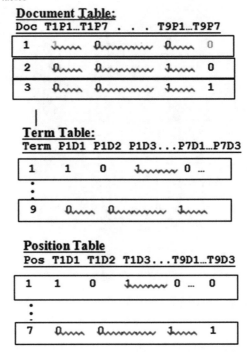

3.1 P-Tree Term View

The term view P-Trees, for example, are then created by bit-slicing the Term Table. They are based on the tweet ID dimension and the position of occurrence of the term within the tweet, a P-Tree for every term is been created (Fig. 4).

3.2 P-Tree Position View

The Position view P-Trees are created by bit-slicing the Position Table. They are based on the tweet ID dimension and the unique term occurrences we find in each tweet ID, a P-Tree for every position is been created (Fig. 5).

Another view which can be created easily is the Document view or tweet ID view by bit-slicing the Document Table. It is based on the unique term and their position in each of the tweets, a P-Tree for every tweet ID can be created.

```
Term - How
```

Positions	0	1	2	3	4	5	6	7	8	9	10
581515351424372736	1	0	0	0	0	0	0	0	0	0	0
581515240791252993	0	0	0	0	0	0	0	0	0	0	0
581509460805464064	0	0	0	0	0	0	0	0	0	0	0
581455867066859520	0	0	0	0	0	0	0	0	0	0	0
581452817216458752	0	0	0	0	0	0	0	0	0	0	0
581449008859729920	0	0	0	0	0	0	0	0	0	0	0
581437140078346240	0	0	0	0	0	0	0	0	0	0	0
581145979891503104	0	0	0	0	0	0	0	0	0	0	0
581137697659600896	0	0	0	0	0	0	0	0	0	0	0
581127573062721536	0	0	0	0	0	0	0	0	0	0	0
581123219668754432	0	0	0	0	0	0	0	0	0	0	0
581115979427659776	0	0	0	0	0	0	0	0	0	0	0

Fig. 4 Term view diagram showing an example of the term view of the term "How"

```
Position - 1
```

Terms	RT	@Forbes:	Risk	&amp;
581980782815797248	0	1	0	0
581899896640372736	0	0	0	0
581515351424372736	0	0	0	0
581515240791252993	0	0	0	0
581509460805464064	0	0	0	0
581455867066859520	0	0	0	0
581452817216458752	0	0	0	0
581449008859729920	0	0	0	0
581437140078346240	0	1	0	0
581145979891503104	0	0	0	0

Fig. 5 Position view diagram showing an example of the position view of the position "1"

The data has been stored redundantly in the form of three different P-Tree views. The computational advantage of processing the data in the form of these bitmaps becomes significant when the data is very deep and also significantly wide. From conventional horizontal processing, we have to throw away data not because it is irrelevant but because the massive amount of data results in data mining times that are too much high. With P-Trees we would not ever need to throw away the data due to the faster processing advantage [9, 11].

4 Classification

4.1 Ticker Symbols

Now with all the terms been stored in the term view P-Trees we can filter on the terms which only begin with a "$" symbol. The term view P-Trees for every user generates the *term frequency* based on the one occurrence of these terms. The ticker symbol which has the highest frequency of one occurrence is the most discussed amongst the investors. This insight is helpful especially when a user needs to know which stocks to invest in primarily and does not have much knowledge on the markets. The system scans through 3000 different twitter user Names and retrieves this information on the most frequent ticker symbol which has been discussed. The system can also process any number of latest tweets as specified to analyze. Once the most frequent occurrence of the ticker symbols is obtained, the system generates an .xml based output containing the ticker symbol frequency and the tweet Text associated with the ticker symbol occurrence.

The analysis has also been done with varied sample sizes of tweets and investors—100 tweets across 100 investors, 10 tweets across 500 investors, and so on. A sample result with the most frequent 10 ticker symbol occurrence from the latest 500 tweets across 25 most tweeted investors between 30 March 2015 and 03 April 2015 is shown in Fig. 6.

We try to see the actual behavior of these stocks in the market within the same time frame the investors were tweeting about these ticker symbols and the results are shown in Fig. 7.

The results show that $TRIL and $LBIO which had the highest frequency of occurrence in that time period had risen close to 36.2% and 12.39%, respectively, in that time period. This is irrespective of studying the sentiment of the tweets and is entirely based on just the frequency of occurrence of the ticker symbols in respective tweets of the investors.

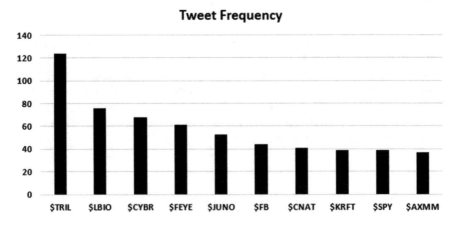

Fig. 6 Top 10 ticker symbols with their respective tweet frequencies

Fig. 7 Top 10 ticker symbols with their respective price changes (%)

Using P-Trees we not only find the most frequently occurring ticker symbols in an efficient manner but also re-construct the tweet texts which were associated for every term and position for the most frequently occurring ticker symbols. We are capturing this information so that the end user can understand the contexts based on which the investors had made the tweets. Some of the sample tweet texts which were associated for the ticker symbol "TWTR" (Twitter Inc.) from different investors are shown below.

```
<TweetWord word="$TWTR" frequency="11">
    <Tweet text="$TWTR Nice breakout to the upside." user-name="Burns277" />
    <Tweet text="$TWTR Nice looking chart to the upside." user-name="Burns277" />
    <Tweet text="$TWTR Ramping" user-name="Burns277" />
    <Tweet text="$TWTR $FB continue their fades" user-name="investorslive"/> </TweetWord>
```

Compared to any horizontal approach such as KNN to scan for the *term frequency*, our approach [12, 13] shows much better results in terms of speed and accuracy. The reason for the improvement in speed is mainly due to the complexity of the two algorithms. Usually, the high cost of KNN based algorithms is mainly associated with their selection phases. The selection phase in our algorithm has a complexity of $O(n)$ where n is the number of dimensions (number of terms) while the KNN approach has a complexity of $O(mn)$ for its selection phase where m is the size of the dataset (number of documents or tuples) and n is the number of dimensions. Drastic improvement is shown when the size of the matrix is very large (the case of $5000 \times 20,000$ matrix size in Table 2). As for accuracy, the KNN approach bases its judgment on the similarity between two document vectors upon the angle between those vectors regardless of the actual distance between them. Our approach does a more sophisticated comparison by using P-Tree AND'ing to compare the closeness of the value of each term in the corresponding vectors, thus being able to judge upon the distance between the two vectors and not only the

Table 2 Time comparison table

Matrix size (document × term)	P-Tree time (ms)	KNN time (ms)
100 × 1000	160	203
500 × 5000	183	332
1000 × 10,000	209	1129
5000 × 20,000	393	13,101

angle. Also, terms that seem to skew the result could have been ignored in our algorithm unlike in the KNN approach which has to include all the terms in the space during the neighbor-selection process.

4.2 Classifying Investors

With the globally most frequent ticker symbols now been classified we would like to evaluate the sentiment of the investors so that we can study their effect on certain ticker symbols. Investors can safely be assumed to be sentiment driven. Since we emphasize so heavily on the tweet texts it is important to make sure that the investors are not underreacting or overreacting for particular ticker symbols. The real investors and markets are very complicated to be summarized by a few selected biases and trading frictions. The *top down* approach [14, 15] focuses on the measurement of reduced form, aggregate sentiment, and traces its effects to market returns and individual stocks. The approach is based on two broad undisputable assumptions of behavioral finance—sentiment and the limits to arbitrage—to explain which stocks are likely to be most affected by sentiment, rather than simply pointing out that the level of stock prices in the aggregate depends on sentiment. In particular, stocks of low capitalization, younger, unprofitable, high volatility, non-dividend paying, growth companies, or stocks of firms in financial distress are likely to be disproportionately sensitive to broad waves of investor sentiment. In particular, stocks that are difficult to arbitrage or to value are most affected by sentiment (Fig. 8).

When sentiment is low, the average future returns of speculative stocks exceed those of bond-like stocks. When sentiment is high, the average future returns of speculative stocks are on average lower than the returns of bond-like stocks.

In our analysis, we also focus on investors being biased on particular ticker symbols and would like to remove any bias [16] in their tweets based on their tweet counts. Moreover, the heterogeneity of the population of investors tweeting on ticker symbol T_s is discussed by a wider population range.

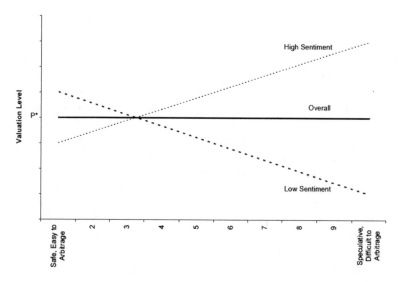

Fig. 8 Cross-sectional effects of investor sentiment. Stocks that are speculative and difficult to value and arbitrage will have higher relative valuations when sentiment is high

5   Sentiment Analysis

The main portion of this research is to run the sentiment analysis on the pulled tweets of selected investors so that we can identify which of the ticker symbols have positive, neutral, and negative sentiment score. Choosing an accurate sentiment analyzer tool is challenging while processing unstructured texts such as tweets [17, 18]. Understanding and analyzing unstructured text is an increasingly popular field and includes a wide spectrum of problems such as sentiment analysis, key phrase extraction, topic modeling/extraction, aspect extraction, and more. Sentiment analysis involves a number of key challenges. One simple approach is to do lexicon based analysis on words or phrases that impart negative or positive sentiment to a sentence the words "*bad*", "*not good*" would belong to the lexicon of negative words, while "*good*", "*great*" would belong to the lexicon of positive words. But this means such lexicons must be manually curated, and even then they are not always accurate. The phrase such as "not bad" which imparts a positive sentiment is hard to detect with simple lexicon based analysis.

A more robust approach is to use machine learning [19–22] to train models that detect sentiment. For training the system a large dataset of text records that was already labeled with sentiment for each record was first obtained. The first step is to tokenize the input text into individual words, then apply stemming. Next we constructed features from these words; these features are used to train a classifier. Upon completion of the training process, the classifier can be used to predict the sentiment of any new piece of text. It is important to construct meaningful features for the classifier, and our list of features includes several from the state-of-the-art research:

- *N-grams* denote all occurrences of n consecutive words in the input text. The precise value of n may vary across scenarios, but it's common to pick $n = 2$ or $n = 3$. With $n = 2$, for the text *"the quick brown fox,"* the following n-grams would be generated—[*"the quick," "quick brown," "brown fox"*]
- *Part-of-speech tagging* is the process of assigning a part-of-speech to each word in the input text. We also compute features based on the presence of emoticons, punctuation, and letter case (upper or lower)
- *Word embeddings* are a recent development in natural language processing, where words or phrases that are syntactically similar are mapped closer together, e.g., in such a mapping, the term *cat* would be mapped closer to the term *dog*, than to the term *car*, since both dogs and cats are animals. Neural networks are a popular choice for constructing such a mapping. For sentiment analysis, we employ neural networks that encode the associated sentiment information as well. The layers of the neural network are then used as features for the classifier.

So the crux of the research involves gathering the sentiment score accurately using a properly trained sentiment analysis tool. In this research we have used Microsoft Azure Sentiment Analyzer [23] to run the sentiment analysis on the pulled tweets.

6 Azure Sentiment Analyzer Comparisons with Other NLP Tools

The classifier was performing well on our internal Microsoft datasets [19, 23, 24], so we wanted to compare it against external offerings. We evaluated its performance against two external services—the Stanford NLP Sentiment Analysis engine (using its pre-trained sentiment model) and a popular commercial tool. Here are the comparative benchmarks.

On datasets comprising tweets, Azure ML Text Analytics was 10–20% better at identifying tweets with positive vs negative sentiment. We used tweets data from Sentiment140 and CrowdScale. Here is the comparison of the three systems on area under the ROC curve (Fig. 9).

On user review datasets, Azure ML Text Analytics was 10–15% better. We analyzed sentiment on a dataset of TripAdvisor reviews, here is a comparison of the results based on the F1 score (Fig. 10).

The Azure Machine Learning Text Analytics outperforms other offerings on short as well as long forms of text for the sentiment analysis task.

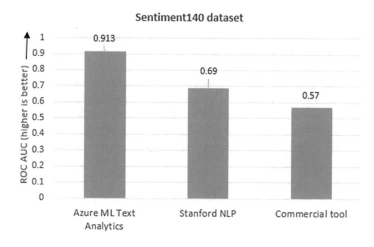

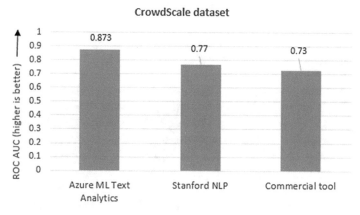

Fig. 9 Azure ML text analytics comparison with other NLP tools on Sentiment140 and Crowd-Scale dataset

7 Sentiment Trend

Our null hypothesis H_0 in the paper is that the sentiment trend of key investors does not bear any relation with the actual stock movement in the market. With the Azure Sentiment Analyzer running on the pulled tweets we can have the sentiment score generated on the pulled tweets for every investor on particular ticker symbol and can observe the trend of the particular ticker over a period of time. A single sentiment score is assigned to the complete tweet. So even if different stock symbols are mentioned in the same tweet the same sentiment score from Microsoft Azure Sentiment Analyzer would be used for each ticker symbol. The Azure sentiment analyzer [19, 22, 23, 25, 26] for every tweet generates a score between 0 and 1 where sentiment score >0.7 denotes high positive sentiment while <0.4 denotes

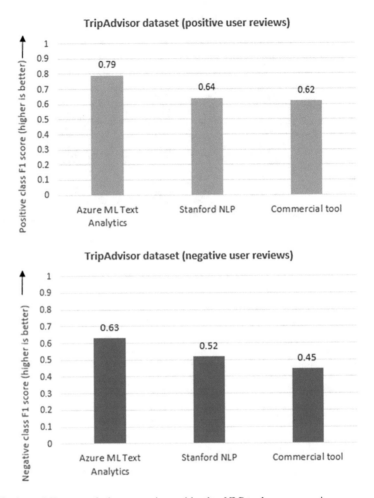

Fig. 10 Azure ML text analytics comparison with other NLP tools on user reviews

low negative sentiment. The sentiment score range between 0.4 and 0.7 denotes the neutral range. We can choose to have various time spans on which the sentiment score has been generated from the extracted tweets and observe how the sentiment line varies. The sentiment analyzer generates a sentiment score trend line over a time span and based on this line strategic investment decisions can be made.

For the month of July, from the pulled tweets of all our focused investors, for every day we were trying to run the Azure sentiment analyzer on the pulled tweets and we calculated the daily average sentiment score. In Fig. 11 we show Daily Avg Sentiment column to be added alongside the fundamentals of the stock symbol for every day. The Daily Avg Sentiment score is the average sentiment score for that day from the sentiment scores for every tweet from every investor. We do similar calculations for all the stocks under Dow Jones Industrial Average and S&P 500.

Ticker Name	Trading Date	Open Price	Close Price	High Price	Low Price	Volume	Daily Avg Sentiment
AAPL	20150701	126.9	126.6	126.94	125.99	30238800	0.578992
AAPL	20150702	126.43	126.44	126.69	125.77	27211000	0.480586
AAPL	20150706	124.94	126	126.23	124.85	28060400	0.590045
AAPL	20150707	125.89	125.69	126.15	123.77	46946800	0.453317
AAPL	20150708	124.48	122.57	124.64	122.54	60490200	0.494714
AAPL	20150709	123.85	120.07	124.06	119.22	77821600	0.486311
AAPL	20150710	121.94	123.28	123.85	121.21	61194200	0.571411
AAPL	20150713	125.03	125.66	125.76	124.32	37237800	0.514514
AAPL	20150714	126.04	125.61	126.37	125.04	31535500	0.554745
AAPL	20150715	125.72	126.82	127.15	125.58	33463100	0.583965
AAPL	20150716	127.74	128.51	128.57	127.35	35866800	0.602476
AAPL	20150717	129.08	129.62	129.62	128.31	45693300	0.497959
AAPL	20150720	130.97	132.07	132.97	130.7	54159800	0.548819
AAPL	20150721	132.85	130.75	132.92	130.32	58898800	0.599529
AAPL	20150722	121.99	125.22	125.5	121.99	1.15E+08	0.542086
AAPL	20150723	126.2	125.16	127.09	125.06	50688700	0.614853
AAPL	20150724	125.32	124.5	125.74	123.9	41051600	0.64991
AAPL	20150727	123.09	122.77	123.61	122.12	44274800	0.568923
AAPL	20150728	123.38	123.38	123.91	122.55	33448900	0.379291
FB	20150701	86.77	86.91	87.95	86.49	25163600	0.750454
FB	20150702	87.4	87.29	87.44	86.34	16719400	0.616716
FB	20150706	86.49	87.55	88.19	86.39	24553200	0.677386
FB	20150707	87.8	87.22	87.85	85.23	33005700	0.603223
FB	20150708	86.29	85.65	86.75	85.45	24311200	0.708868
FB	20150709	86.73	85.88	87.6	85.65	23144200	0.732593
FB	20150710	87.35	87.95	88.22	86.77	23148600	0.718135
FB	20150713	88.66	90.1	90.22	88.42	29579200	0.614639
FB	20150714	90.46	89.68	90.8	89.65	26516100	0.658844
FB	20150715	90	89.76	90.99	89.42	30766400	0.589445
FB	20150716	90.28	90.85	90.86	89.77	21020700	0.651392
FB	20150717	92.55	94.97	95.39	92.54	53482200	0.661951
FB	20150720	95.85	97.91	98.6	95.36	48510100	0.626412
FB	20150721	98.95	98.39	99.24	97.14	38848800	0.640798
FB	20150722	96.74	97.04	97.58	95.92	28083900	0.632358
FB	20150723	96.96	95.44	97.45	94.81	28838800	0.630107
FB	20150724	97.35	96.95	97.76	95.88	33229500	0.692368
FB	20150727	96.58	94.17	96.61	93.83	38285100	0.609159
FB	20150728	94.84	95.29	95.56	93.31	34180500	0.645587

Fig. 11 Daily average sentiment score calculated on pulled tweets using azure sentiment analyzer for the month of July for companies Apple and Facebook

Once we have this information we would like to analyze the sentiment score over a particular time span and if it is reflective of the actual performance of the stock in the market. In Fig. 12 we show how the Daily Avg Sentiment line varies in comparison to the closing price of the stock over a one month time span.

We can see from the above figures that when the daily average sentiment trend is showing upward trends then the stock either shows pretty much stable or moves in a positive direction and when the sentiment trend is downwards then the stock price movement also shows a downward movement thus resulting in results to reject the null hypothesis.

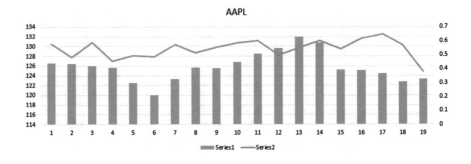

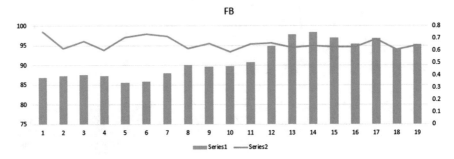

Fig. 12 Daily average sentiment score in comparison to closing price of stocks on pulled tweets using azure sentiment analyzer for the month of July for companies Apple and Facebook

8   Future Direction

With the twitter data now being accessible and loaded in P-Tree views, the future potential in this area can be significant. P-Trees will help us analyze if necessary trillion records deep. In our approach we believe that adding the third dimension, namely term position in the tweet, offers us a wide range of possible new decision supports such as phrase analysis (multiple juxtapose terms). For example, the evaluation of "Buy AAPL," could be done by shifting the "AAPL" P-Tree with one position bit, then by using AND operator with the "Buy" P-Tree. In addition to phrase analysis, the position dimension also opens up the potential of discovering signals based on where in a tweet, a particular term appears (early on, later, etc.). Of course, introducing the third dimension of *term position* massively expands the size of the data cube being considered and necessitates effective compression techniques. Fortunately, when converting to P-Trees in each of the three dimensions, it is true that in all but one of the dimensions (document dimension) each P-Tree is as sparse as it could possibly be (a single one-bit) and therefore compresses maximally [11, 27]. In the document dimension, by replicating single one-bit P-Trees, instead of using links, the same advantage can be realized. These advantages will be pursued in the future.

The main idea illustrated in this paper is solving a classification problem to understand which ticker symbols we would focus on depending on the tweets made by the investors. Some users can also trade against this information based on their investment pattern. The actual movement of the stocks in the market in various time frames and their fluctuations can be related with the information impact from this data. However, we are not classifying investors in our approach and we would like to consider doing so in the future, where based on their tweet ticker symbol references we can observe the actual behavior of those stocks in the market and weigh the investors accordingly.

We also show a very simple method to compute the average sentiment score from the pulled tweets of investors and we are not weighing the tweets based on the relevance of when it is being tweeted. An alternative approach would be to calculate the weighted sentiment score where weights would be assigned based on the relevance of when the tweet was made. Recent tweets from key investors would have higher weights assigned to the sentiment score.

We are also extensively exploring Clique Search algorithm and other clustering techniques to group similar stocks together and define better strategies based on which investment decisions could be made.

We also plan to include this solution as a phone app which will be commercially available to the end consumers on Windows, Android, and the iOS platforms.

9 Conclusion and Limitations

We have shown in this paper how using Twitter API and P-Trees we can provide insight to users who would like to invest in the market but cannot identify which ticker symbols they would like to invest in. They can use this solution which can provide them insight to make business investment decisions. We have also classified the investors based on their tweet counts and have ensured that any bias from the measures is removed. Sentiment analysis is gradually becoming an important aspect of investment when users can understand the social sentiment score of a particular ticker symbol upfront from the discussions made by key investors and then based on the obtained information they can decide where to invest. In this paper we did not discuss about any particular strategies we follow, but as we are gathering more data every day we are actively researching multiple algorithms to provide us with a much more comprehensive decision making ability.

References

1. Ding Q, Khan M, Roy A, Perrizo W. The P-tree algebra. Proceedings of the ACM SAC, Symposium on Applied Computing. Madrid; 2002.
2. Khan M, Ding Q, Perrizo W. K-nearest neighbor classification on spatial data stream using P-trees. Proceedings of the Pacific-Asia Conference on Knowledge Discovery and Data Mining (PAKDD 02). Taipei; May 2002. pp. 517–28.

3. Perrizo W, Ding Q, Roy A. Deriving high confidence rules from spatial data using peano count trees. Proceedings of the WAIM, International Conference on Web-Age Information Management. Xi'an; July 2001. 91–102.
4. Rahal I, Perrizo W. Query acceleration in multi-level secure database systems using the P-tree technology. Proceedings of the ISCA CATA, International Conference on Computers and Their Applications. Honolulu, Hawaii; March 2003.
5. Han J, Kamber M. Data mining: concepts and techniques. San Francisco,CA: Morgan Kaufmann Publishers Inc.; 2000.
6. Adriaans P, Zantinge D. Data mining. Addison Wesley; 1996.
7. Salton G, Buckley C. Term-weighting approaches in automatic text retrieval. Inf Process Manag. 1988;24(5):513–23.
8. Salton G, Wong A, Yang CS. A vector space model for automatic indexing. Commun ACM. 1975;18(11):613–20.
9. Rahal I, Perrizo W. An optimized approach for KNN text categorization using P-trees. Proceedings of the 2004 ACM Symposium on Applied Computing (SAC-04) Nicosia; 14–17 March 2004.
10. Rahal I, Ren D, Perrizo W. A scalable vertical model for mining association rules. J Inf Knowl Manag (JIKM). 2004; 3(4): 317–29.
11. Perera A, Abidin T, Serazi M, Hamer G, Perrizo W. Vertical set squared distance based clustering without prior knowledge of K. International Conference on Intelligent and Adaptive Systems and Software Engineering (IASSE-05), Toronto; 20–22 July 2005. pp. 72–7.
12. Abidin T, Perrizo W. SMART-TV: a fast and scalable nearest neighbor based classifier for data mining. Proceedings of the 21st Association of Computing Machinery Symposium on Applied Computing (SAC-06), Dijon; 23–27 April 2006.
13. Rahal I, Serazi M, Perera A, et al. DataMIME™. Association of Computing Machinery, Management of Data (ACM SIGMOD 04), Paris; June 2004.
14. Baker M, Wurgler J. ©Investor sentiment and the cross-section of stock returns. J Finance. 2006;61:1645–80.
15. Baker M, Wurgler J. ©Investor sentiment in the stock market. J Econ Perspect. 2007;21:129–51.
16. Yang C, Fayyad UM, Bradley PS. Efficient discovery of error-tolerant frequent item sets in high dimensions. Proceedings of the KDD; 2001. pp. 194–203.
17. Shuai X, Pepe A, Bollen J. How the scientific community reacts to newly submitted preprints: article downloads, twitter mentions, and citations. PLoS ONE. 2012;7(11):e47523. doi:10.1371/journal.pone.0047523.
18. Li D, Ding Y, Shuai X, et al. Adding community and dynamic to topic models. J Informet, 6(2): 237–253, 2012
19. Introducing Text Analytics in the Azure ML Marketplace by NagenderParimi. http://blogs.technet.com/b/machinelearning/archive/2015/04/08/introducing-text-analytics-in-the-azure-ml-marketplace.aspx.
20. Bollen J, Mao H. Twitter mood as a stock market predictor. IEEE Comput. 2011;44(10):91–4.
21. Bollen J, Gonçalves B, Ruan G, Mao H. Happiness is assortative in online social networks. Artif Life. 2011;Summer-17(3):237–51.
22. Bollen J, Mao H, Zeng X. Twitter mood predicts the stock market. J Comput Sci. 2011; 2(1): 1–8. doi:10.1016/j.jocs.2010.12.007, arxiv: abs/1010.3003. (featured on CNBC, CNN International and Bloomberg News!).
23. Azure Machine Learning group at Microsoft Research and Development. https://azure.microsoft.com/en-us/services/machine-learning/.
24. Bollen J, Pepe A, Mao H. Modeling public mood and emotion: twitter sentiment and socio-economic phenomena ICWSM11, Barcelona, July 2011 (arXiv: 0911.1583)—poster.
25. Mao H, Pepe A, Bollen J. Structure and evolution of mood contagion in the twitter social network. Proceedings of the International Sunbelt Social Network Conference XXX, Riva del Garda, July 2010. (abstract).

26. Mao H, Counts S, Bollen J Quantifying the effects of online bullishness on international financial markets. European Central Bank Statistics Papers Series. 2015.
27. Ren D, Wang B, Perrizo W. RDF: a density-based outlier detection method using vertical data representation. Proceedings of the 4th Institute of Electrical and Electronic Engineers (IEEE) International Conference on Data Mining (ICDM-04); 1–4 Nov 2004. pp. 503–6.

Mining Community Structure with Node Embeddings

Thuy Vu and D. Stott Parker

1 Introduction

Mining of network data has gained great importance in recent years due to the significant developments in online social networking services and the wide applicability in different academic disciplines [6, 8]. After decades of development on better mining algorithms [10], scalability has become the decisive factor in recent data science efforts. For example, network approximation has been explored as one solution for computational infeasibility in large network analysis problems, including detecting communities using betweenness in networks of several thousands of nodes (or more) [23] and influence propagation in large social networks [9].

Recently, the community in artificial intelligence (AI) and machine learning (ML) has made great progress in training representations using large artificial neural networks in a reasonable amount of time [12]. This has enabled many applications and also spawned new research in machine learning areas involving *deep learning* using neural networks. Distributed representation learning using recurrent neural networks is one successful beneficiary of this research. Techniques in distributed representation learning have almost instantly set many new state-of-the-art records. More impressively, recent advances have shortened neural network training times for large data to reasonable levels. One recent example is the reduction of learning time for word representations training on 6 billion words to one day with roughly 300 cores [16], a decrease from 2500 cores in 2011 [5] (more than a factor of 8) without significant compromise in performance.

T. Vu (✉) • D.S. Parker
Department of Computer Science, University of California, Los Angeles,
Los Angeles, CA 90095, USA
e-mail: thuy@cs.ucla.edu; stott@cs.ucla.edu

© Springer International Publishing AG 2017
M. Kaya et al. (eds.), *From Social Data Mining and Analysis to Prediction and Community Detection*, Lecture Notes in Social Networks,
DOI 10.1007/978-3-319-51367-6_6

Adoption of recurrent neural networks for distributed representation learning began early and is still gaining attention in many fields in computer science, including computer vision, speech recognition, and natural language processing (NLP). In NLP, deep learning has been adopted rapidly and in diverse ways to find better distributed representation for words, also known as word embeddings. Word embeddings have been successfully integrated in many NLP problems, in areas such as part-of-speech tagging and machine translation [5], dependency parsing [3, 13], and even sentiment analysis [26] with impressive results. Deep representation learning is also studied in speech recognition [11] and image processing [15].

To the best of our understanding, few attempts have been made to apply neural networks for distributed representation learning in social network analysis and mining. Previously, we have presented applications of node embeddings in community-based network analysis [30]. In this paper, we extend our previous work in analyzing applications of node embeddings to the community detection problems. Our contribution is threefold:

1. We introduce a simple yet generic method to learn embeddings for nodes in a network. Specifically, we adapt Skip-gram context handling algorithm for words in natural language processing to learn embeddings for nodes based on both their connectivity in the network and their own attributes.
2. We propose to use the learned embeddings to re-examine links in social networks. In particular, we employ embedding-based similarity metrics between nodes to re-adjust networks—adding new potential links and discarding distance-based irrelevant links. We experimentally show that newly re-adjusted networks can be beneficial in standard link-based community detection algorithms. We hypothesize that many algorithms on networks can do a better job when nodes are better spaced and links are more relevant.
3. We also show how node embeddings can be easily considered for mining tasks that require similarity computing in networks, including community homogeneity, distance, and identification of *community connectors* (nodes with the important property of connecting communities). By treating research disciplines as communities, many interesting findings about the computer science literature have emerged from this work.

In addition, we construct ground truth for community detection in the computer science literature, DBLP, including best paper annotation in different conferences over the years for our experimental evaluations. We position our research with respect to distributed representation learning and community-based mining in social networks in Sect. 2. We then present our proposed node embeddings training in Sect. 3. We show how link re-adjustment and community-based mining are beneficial from the learned embeddings in social networks in Sects. 4 and 5, respectively. Our empirical studies are presented in Sect. 6, followed by conclusions in Sect. 7.

2 Background

2.1 Representation Learning

This paper is related to recent studies in learning distributed representations, especially in natural language processing (NLP). Distributed representations of concepts, known as embeddings, are usually encoded by high dimensional real-valued vectors. In general, the embeddings are learned through optimization of an objective function automatically on large volumes of unannotated data.

In NLP, neural-based language models represent language by embeddings of words in a high dimensional vector space. The learned word embeddings are then introduced for certain tasks that require word-based comparisons. Mikolov et. al. proposed continuous bag-of-words (CBOW) and Skip-gram, the current state-of-the-art models [16]. Both are implemented and available from the word2vec project page: code.google.com/p/word2vec/. The main difference between these two models is in whether the objective is to learn representations to predict words (CBOW) or contexts (Skip-gram).

Our node embedding learning model is adapted from word2vec. We consider Skip-gram for its sensitivity in dealing with low connection nodes, which is crucial when dealing with large social networks. Our work is also related to other research in learning neural-based language models [5], neural word embeddings [13, 14], word embeddings for speech recognition [1], and feature embeddings for dependency parsing [3].

2.2 Social Network Analysis and Mining

In recent years, the development of online social services has brought significant attention and rapid advances to studies in social network analysis and mining. Among different social network applications, community detection remains important because structural information is fundamental for large scale network analysis and mining. Community detection in networks has been extensively studied in the literature. Algorithms for this problem can be classified in three approaches with respect to their relying resources.

The first and most popular approach is to maximize distinctions in density among communities using network connectivity. For example, the algorithm proposed in [20] maximizes modularity score for a community setting of the network using edge-betweenness centrality. Even though this is a reliable method for best performance—aiming to maximize the evaluation metrics directly—it does not scale for networks with thousands of nodes due to high computation cost in repeated recomputation of edge betweenness after each iteration.

The second approach considers node-based features to determine communities. One exemplary work for this approach is [29], which developed the SNAP algorithm for grouping nodes using their attributes and relations in node connections. Node relations are among the identifying features for nodes, but are not relevant for the density measures considered in the previous approach. It is, however, worth noting that this approach performs best with sparse networks, where the network's degree distribution is largely uniform and the density is flat. It may not be effective for a regular social network when degree distributions follow a power law [4].

The third approach is a hybrid of the previous two. Existing works on this approach include [25, 28, 31, 32]. In this paper, we suggest a strategy to bridge methods from the aforementioned two approaches. Thus, our proposed method should be classified in this category. Indeed, our node embeddings are basically learned from node descriptions, including attributes and relations with neighbors. And algorithms in the first approach can perform density-based community detection with similarity scores derived from the learned embeddings.

3   Distributed Representation in Networks

We present our proposed adaptation of the Skip-gram model for node embeddings training in this section. In general, we consider node attributes and links as *context* in a tuple (*node, context*) to describe nodes of a network.

Formally, let $G(V, E, \mathscr{A})$ be a network with set of nodes V and set of edges E. Edges in E can be of arbitrary annotated type in a heterogeneous information network [27]. In addition, every node $v_i \in V$ is associated with descriptive attributes $\alpha_i \in \mathscr{A}, \alpha_i = \{a_1; a_2; \cdots a_{|\alpha_i|}\}$. Let D be the training input describing V, denoted as $\{(v_i, c_i)\}_{i=1}^{|V|}$, such that $c_i = (\alpha_i, \{v_j : (v_i, v_j) \in E\})$. The objective is as follows:

$$\max_{\mathbf{x}_v, \mathbf{x}_c} \left(\sum_{(v,c) \in D} \log \sigma(\mathbf{x}_c \cdot \mathbf{x}_v) + \sum_{(v,c) \in D'} \log \sigma(-\mathbf{x}_c \cdot \mathbf{x}_v) \right)$$

where $\sigma(x) = 1/(1 + e^{-x})$ and $\mathbf{x}_v \in \mathbb{R}^d$ is the embedding of node v. D' is generated from D through a negative sampling process. Specifically, $(v_i, c_i) \in D'$ means there exist indices x and y such that $(v_i, c_x | x \neq i) \in D$, $(v_y, c_i | y \neq i) \in D$, and $(v_i, c_i) \notin D$.

The training process maximizes the objective via a stochastic-gradient process. The output of the training includes embedding $\mathbf{x}_{v_i}$ for each v_i and embedding $\mathbf{x}_{c_j}$ for each context c_j. These embeddings for v_i and c_j should be proximally close if $(v_i, c_j) \in D$. As a result, two nodes sharing the same context should also be in close proximity to each other. The proximity between nodes can be estimated using Euclidean distance between their embeddings.

4 Embedding-Based Community Detection (EBCD)

Two factors that define community are the connectivity density and the topical similarity of the vertices in the community. In the literature, many algorithms focus on the former through maximizing the community density by grouping vertices to maximize some graphical metrics, usually modularity. However, this might not yield satisfactory results in some applications when vertex similarity is more important. In addition, density-based algorithms usually suffer from computational infeasibility as the network size increases [23].

The immediate utility of node embeddings is the vector-based comparison among nodes in the network. This comparison has been proven effective in NLP literature as we introduced earlier. Moreover, training of the embeddings can scale well and can be performed in an online setting. This benefits in handling online data in social network. Finally, a wide range of descriptive features can be embedded in the same vector space, which makes comparisons feasible for networks of different types.

We consider three different ways to incorporate this similarity in community detection algorithms.

4.1 Node Clustering Algorithms

We consider nodes to be located independently in a d-dimensional space and assume no connections between nodes. We basically remove all edges and apply a vector quantization-based clustering algorithm to group vertices into K different groups depend on their proximity. Many clustering algorithms can fit this setting. For example, K-means and its variations including K-medians and K-medoids can be a good fit, depending on the nature of data. We consider K-means our base clustering algorithm in this paper. The basic idea of this algorithm is to partition a set of n observed data entities into K different clusters using proximity scores.

Even though the algorithm itself has some merit and the embeddings are proven representations, it is unusual for this method to yield impressive result. The biggest problem is its lack of consideration for network connectivity.

4.2 Community Detection with Weighted Network

We consider incorporating the distances between nodes directly into the network—as network weights. The literature shows that most community detection algorithms can handle weighted networks [7]. If an algorithm does not, however, it is still possible to consider weights by converting the network into unweighted multigraph format [17]. This conversion allows multiple edges between two nodes.

Since the distance between embeddings means dissimilarity ($t \geq 0$), it is usually necessary to convert it into similarity ($s \geq 0$) network before passing into some certain algorithms. There are many ways to convert between two metrics. Some standard conversion methods include:

- $s = 1 - t$, this conversion is only applicable when $t \in [0; 1]$
- $s = t^{-1}$, similarly, this conversion is straightforward and requires $t > 0$
- $s = (1 + t)^{-1}$, this conversion is similar to $s = t^{-1}$ without conditioning on t
- $s = e^{\frac{-\beta \cdot t}{std(t)}}$, this is known as the heat kernel and can be simplified as follows:
- $s = e^{-t^2}$

Adding weights into the network does encourage the algorithms to group closer nodes into the same community. However, this modification might not result in big differences in outcome since the algorithm itself still considers dense connection its main objective function via link structure. In other words, if the connectivity structure of the network remains unchanged, little difference in outcome can be expected.

4.3   (α; β) LInk Re-Adjustment in Networks (LIRA)

This strategy aims at resolving both of the above-mentioned weaknesses through combination. The direct clustering strategy has its own merits in combining nodes through their similarities. However, simply embedding distances in the network shows little effect when the network is sparse. Therefore, we propose to use grouping results from node clustering algorithms to re-adjust connectivity in the network. Specifically, we consider two operations on network edges:

- *disjoin* two nodes, removing the in-between edge, if they belong to two distant groups given by a clustering algorithm. The threshold $\alpha\%$ means that nodes belonging to the top $\alpha\%$ most-distant communities should be disjoined. Distance between communities is computed using the method in Sect. 5.2.
- *join* two nodes if they belong to clusters that are immediately adjacent to each other. Each node is joined with the top $\beta\%$ such nodes.

4.4   Embedding-Based Community Detection

Algorithm 1 presents our proposed integration of the network pre-processed by LIRA with a standard community detection algorithm, namely Embedding-based Community Detection (EBCD). This integration is generic and makes it easy for users to consider their favorite algorithms for clustering (as in line 2) and community detection (as in line 5).

Algorithm 1 Embedding-based community detection

 1: **procedure** EBCD($G(V, E), \mathbf{e}_V, \alpha, \beta$)
 2:     $K \leftarrow$ CLUSTERING($\mathbf{e}_V$)   ▷ e.g. K-means
 3:     $(D, J) \leftarrow$ LIRA(G, K, α, β)
 4:     $G' \leftarrow G(V, E - D + J)$
 5:     $C \leftarrow$ COMDET(G')   ▷ a generic community detection algorithm
 6:     **return** C ▷ community assignments
 7: **end procedure**
 8:
 9: **procedure** LIRA($G(V, E), K, \alpha, \beta$)
10:     $D \leftarrow \emptyset$
11:     **for all** $e \in E$ **do**
12:         $(v_f, v_g) \leftarrow e$ ▷ get both ends of e
13:         $c_{v_f} \leftarrow K(v_f)$ ▷ get cluster of v_f
14:         $c_{v_g} \leftarrow K(v_g)$ ▷ get cluster of v_g
15:         **if** $c_{v_f \text{ or } v_g} \in$ the α % most distant to $c_{v_g \text{ or } v_f}$ **then**
16:             $D \leftarrow D + e$ ▷ include e into D
17:         **else if** $c_{v_f \text{ or } v_g} \in$ the β % most related to $c_{v_g \text{ or } v_f}$ **then**
18:             $J \leftarrow J + e$ ▷ include e into J
19:         **end if**
20:     **end for**
21:     **return** (D, J)
22: **end procedure**

Although LIRA networks are used in EBCD in this paper, we believe EBCD is also suited for other kinds of network analysis applications.

5 Mining in Social Networks

We present three applications of node embeddings in this section. First, we analyze homogeneity of communities in a network to study their concentration. Second, we propose to compute distance between communities. Last, we identify connectors of communities in a network. All computations for these applications rely on node embeddings to derive node similarities.

5.1 Community Homogeneity

Homogeneity for a community represents the degree of closeness of its members. Indeed, distances between members in a community reflect not only their topological structure but also their attributed dissimilarity. We define the homogeneity $h(\cdot)$ of a community c by the variance of all mutual distances:

$$h(c) = \left[\sum_{i=1}^{|E_c|} p_i \cdot (d_i - \mu_c)^2 \right]^{1/2}$$

Here $\mu_c = \sum_{i=1}^{|E_c|} p_i \cdot d_i$ is the mean of all distances in the community.

5.2 Community Distance

There are many ways to define distance between two groups of entities. Distances in this paper consider all possible connections between two communities. Specifically, our distance between communities is the average total distance of all members. Formally, the distance $d(\cdot, \cdot)$ between two communities $\mathbf{C}_i$ and $\mathbf{C}_j$ is defined as:

$$d(\mathbf{C}_i; \mathbf{C}_j) = \frac{\sum_{(u;v) \in \mathbf{C}_i \times \mathbf{C}_j} d(u; v)}{|(u; v) \in \mathbf{C}_i \times \mathbf{C}_j|}$$

where u and v belong to community $\mathbf{C}_i$ and $\mathbf{C}_j$, respectively.

5.3 Community Connectors Identification

Finally, we define a *connector of a community* $\mathbf{C}_i$ *to community* $\mathbf{C}_j$ as a node $v \in \mathbf{C}_i$ that is closest (having smallest average distance) to all nodes in $\mathbf{C}_j$. This is a type of outlier that sits between communities. In other words, the set of connectors is:

$$c(\mathbf{C}_i, \mathbf{C}_j) = \{v \in \mathbf{C}_i : d(\{v\}; \mathbf{C}_j) = \min_{k \in \mathbf{C}_i} d(\{k\}, \mathbf{C}_j)\}$$

Unlike influencers or leaders, community connectors are a type of inter-community outlier, and thus, not necessarily well known. They usually play significant roles in a social network, especially in increasing communication among communities.

6 Experiments

We present our experimental studies for our node embeddings in this section.

6.1 Dataset Construction

We consider the DBLP citation network compiled by aminer.org in our experiments. We downloaded the version released in September 2013 (arnetminer. org/lab-datasets/citation/DBLP_citation_Sep_2013.rar). This dataset consists of 2,244,018 papers and 2,083,983 citation relationships for 299,565 papers (about seven citations each). Each paper has title, authors, publishing venue, abstract, and citations.

We classify all papers and authors into fields in computer science. We consider 24 research fields compiled by Microsoft Academic Search (academic. research.microsoft.com). In each field, a list of related conferences, or publishing venues, is provided. We have a total of 2695 different publishing venues in computer science classified into 24 different research fields.

It is not straightforward to align these research fields to aminer.org's 8881 publishing venues. We assign research fields for each of these venues using text classification on titles, and use these to build our community dataset. Specifically, we follow two steps:

1. We leverage clean alignments (exact matching) between two lists for training data for title classifiers. Altogether 798,293 papers in 2123 different publishing venues have a cleanly assigned field ($\sim$81.8%). We use this as training set for string classification of titles—training 24 Naive Bayes classifiers for 24 fields using the Natural Language Toolkit (nltk.org). These classifiers share the same feature set of the 4572 most frequent words, compiled by unifying the top 1000 most frequent words in each field (with English stopwords removed). This vector size is large and representative for our classification tasks. Finally, each classifier is trained with the positive samples against a random set of negative samples having a size of five times the number of positive samples to sharpen its discrimination against false classifications.
2. We assign fields to a venue by identifying the most popular field assigned to its papers though plurality voting. In addition, fields for 1,431,652 unclassified papers are assigned via their venues. Similarly, voting is also used to assign authors to fields as members. An author can be a member of a single field, while still being an author in multiple fields. There are 1,260,485 authors assigned memberships in 24 research fields in our dataset.

In addition, we also tag best papers in this dataset. We use the list of best paper awards since 1996 consolidated by Jeff Huang (jeffhuang.com/best_paper_awards. html). Table 1 displays the statistics for our annotated community dataset. This dataset is useful for data mining studies on Computer Science research.

Table 1 Statistical results in computer science

Computer science field	#Member	#Author	#Paper
Algorithms & theory	134,219	984,720	433,250
Artificial intelligence	42,623	272,911	100,167
Bioinformatics & computational biology	32,896	126,921	43,481
Computer education	9177	40,551	15,160
Computer vision	21,539	153,613	68,446
Data mining	4294	37,101	12,909
Databases	25,610	162,345	59,139
Distributed & parallel computing	22,338	161,578	53,603
Graphics	11,809	64,061	21,493
Hardware & architecture	34,752	188,350	61,125
Human–computer interaction	209,087	748,969	262,602
Information retrieval	14,713	81,586	26,056
Machine learning & recognition	19,675	117,681	40,575
Multimedia	19,638	110,776	35,842
Natural language & speech	26,989	149,874	53,617
Networks & communications	257,752	1,220,809	431,104
Operating systems	1261	9906	3121
Programming languages	14,655	93,974	48,572
Real-Time & embedded systems	279,306	946,254	328,979
Scientific computing	4922	18,189	7066
Security & privacy	16,560	82,131	31,265
Simulation	3635	14,348	4995
Software engineering	38,863	173,551	63,863
World wide web	14,172	68,034	23,515
Total	1,260,485	6,028,233	2,229,945

6.2   Citation-Based Author Embeddings in DBLP

We study the citation networks of authors in different fields in DBLP. For each paper, we extract two lists of authors: the paper authors (citers) and the authors of citing papers (citees). All possible connections between two author lists are extracted and passed to the representation learning in order to learn citation-based author embeddings. Embeddings for each author are the outcome of the learning process. More importantly, the proximity between authors denotes the similarity of their citing behavior. Presumably researchers working in a close community should exhibit high similarity (low distance) in citing behavior.

Each citation-based author embedding is a vector of 200 real numbers. We also filter out authors having a total of fewer than 500 citations in all their papers.

One immediate result of the learned embeddings is the ability to query *most similar* embeddings using Euclidean distance. In particular, our embeddings reflect citations from researchers to researchers, so *most similar* should mean high overlap

Table 2 Most similar researcher query

Top	"e(Christos Faloutsos)"	Field	Distance
1	Hanan Samet	Algorithms	0.958
2	Caetano Traina Jr.	Databases	0.954
3	Thomas Seidl	Databases	0.953
4	Hans-Peter Kriegel	Databases	0.952
5	Christian Böhm	Databases	0.952
	"e(Christos Faloutsos)\e(Philip S. Yu)"		
1	Hans-Peter Kriegel	Databases	0.353
2	Edwin R. Hancock	Vision	0.334
3	David A. Forsyth	Vision	0.327
4	Thomas S. Huang	Vision	0.326
5	Alberto Del Bimbo	Networks	0.323

in citing behavior. This shows high efficiency of the embeddings since the size is limited to 200 instead of the total number of researchers in the naive strategy. In addition, the embeddings can also capture other different aspects of citation activities. This provides us a tool for similarity queries. We denote $e(.)$ the embedding representation of an actor. Table 2 shows a query example of the five researchers most similar to Christos Faloutsos, who has reportedly made 5239 citations to 5060 different researchers as of 2013. Christos Faloutsos is also one of the leading scientists in the "Data Mining" field. The first half of Table 2 shows that his citations largely overlap with many other leading researchers in "Databases." This is not a surprise since these two fields are historically related. In the second half of Table 2, we display Christos Faloutsos's top five most similar researchers after *subtracting* the embedding of Philip S. Yu, another famous researcher in "Data Mining." Interestingly, this reveals that Christos Faloutsos shares significant interests with researchers in Computer Vision, especially in content-based retrieval.

In the following subsections, we continue to present experimental results using the resulting embeddings for other mining tasks.

6.3 Community Detection Results

In this section, we present the evaluation of our proposed strategy for community detection, EBCD. We consider K-means for clustering and five different algorithms for community detection: Fast Greedy [18], Walk Trap [22], Leading Eigenvector [19], InfoMap [24], and Multi-level [2].

Our evaluation metrics include:

- Variation Inf.—variation of information; lower is better.
- Normalized MI—normalized mutual information; higher is better.

Table 3 Community detection result comparison

Algorithm	Metric	*Baseline*	*Weighted*	$(0.2; 0.2)$	$(0.7; 0.2)$
Fast greedy, 2003	Variation inf.	2.611	2.696	**2.552**	2.598
$O((E + V) \times V)$ [18]	Normalized MI	0.190	0.239	0.256	**0.267**
	Split-join	**5018**	5394	4788	5061
	Adjusted rand	0.086	0.098	0.123	**0.133**
	Modularity	0.287	**0.366**	0.312	0.345
Walk trap, 2005	Variation inf.	3.049	3.162	3.207	**3.045**
$O(E \times V^2)$ [22]	Normalized MI	0.402	0.409	0.407	**0.415**
	Split-join	**5337**	5466	5398	5393
	Adjusted rand	0.210	**0.211**	0.204	**0.211**
	Modularity	**0.297**	0.268	0.250	0.249
Leading eigenvector, 2006	Variation inf.	2.934	2.721	**2.765**	2.740
$O((E + V) \times V)$ [19]	Normalized MI	0.214	0.240	0.202	**0.317**
	Split-join	5813	5575	5073	**4908**
	Adjusted rand	0.082	0.094	0.073	**0.115**
	Modularity	0.320	**0.329**	0.302	0.306
InfoMap, 2008	Variation inf.	2.633	2.575	2.513	**2.405**
$O(E)$ [24]	Normalized MI	0.380	0.385	0.401	**0.403**
	Split-join	4831	4744	4618	**4598**
	Adjusted rand	0.196	0.202	0.205	**0.207**
	Modularity	0.419	**0.428**	0.393	0.396
Multi-level, 2008	Variation inf.	2.704	2.675	**2.669**	2.671
$O(V \times \log V)$ [2]	Normalized MI	0.316	**0.348**	0.333	0.333
	Split-join	5101	5014	5010	**5009**
	Adjusted rand	0.168	0.182	**0.193**	0.188
	Modularity	0.419	**0.423**	0.399	0.403

- Split-Join—node-based edit distance between two settings; lower is better.
- Adjusted Rand—Rand considers grouping chance; higher is better.
- Modularity—division strength of a network into modules; higher is better.

Tables 3 and 4 present our results. We consider standard outputs of these community detection algorithms on original unweighted networks as baselines, presented in the first column *Baseline* in Table 3. The next column, *Weighted*, shows the results on networks with edges weighted by the embedding distance of two end nodes. Underlined results denote better scores between *Baseline* and *Weighted*. As seen, results on weighted networks are mostly better than the baseline's, winning 21/25 comparisons.

The following two columns present the results of EBCD in two different parameter settings of (α, β) for LIRA. The second setting ($\alpha = 0.7$ and $\beta = 0.2$) is harsher than the first one ($\alpha = 0.2$ and $\beta = 0.2$) since a large amount of edges are filtered out. Specifically, of the 724,301 links in the original network,

Table 4 Community detection result of InfoMap

Metric	Baseline	Weighted	K-means	EBCD(0.7; 0.2)	Truth
VI	2.633	2.575	4.844	**2.405**	0.000
NMI	0.380	0.385	0.081	**0.403**	1.000
Split-join	4831	4744	8225	**4598**	0.000
Adjusted rand	0.196	0.202	0.019	**0.207**	1.000
Modularity	0.419	**0.428**	0.146	0.396	0.381

257,559 ($\sim$35.56%) and 109,854 ($\sim$15.17%) links are filtered out with $\alpha = 0.7$ and $\alpha = 0.2$, respectively. After that, another 24,986 ($\sim$3.4%) and 27,650 ($\sim$3.8%) links are introduced with $\alpha = 0.2$ for these two settings.

As seen, results of EBCD outperform *Baseline* and *Weighted* in 16/25 comparisons, which implies that applying LIRA on networks as preprocessing for community detections can not only simplify the network structure thus resulting in faster computation, but also improve the overall performance when many distance-based irrelevant links are filtered out. Among five algorithms are considered, Walk Trap, Leading Eigenvector, and InfoMap usually yield best performance, which is in accordance with previous benchmarks [7, 21]. Finally, none of EBCD results could produce better performance in modularity comparisons. In other words, high modularity is only recorded when algorithms consider the full network, in either the original or our weighted networks. There are multiple reasons for this outcome; however, we think this is reasonable considering the modularity score of 0.381 of the ground truth, shown in Table 4.

In Table 4, we compare community detection results of InfoMap with K-means, by ignoring all link information (as presented in Sect. 4.1), and the ground truth. The results support our hypothesis that even though we cannot discard all links in networks for community detections, we can apply adjustments on links for better results. In addition, the results in Tables 3 and 4 indicate that simply maximizing network modularity might not be the best objective function for real social networks.

6.4 Mining in Community Data Results

Table 5 shows the *homogeneity* score for inbound citations to a research community (IC), and outbound citations to communities outside (OOC). The IC can be interpreted as the inner diversity in a community. We record comparable high homogeneity OOC scores for most communities. This shows that mutual communication among the fields is highly diverse. The result also indicates that *Machine Learning & Pattern Recognition* community is very selective in citing papers from outside. At the other extreme, it is interesting that though sitting at different extremes in IC score ranking, *Scientific Computing* and *Hardware & Architecture* are the top two fields in citing outsiders. This might be explained by their research results

Table 5 Homogeneity analysis: in-community (IC) and out-of-community (OOC), sorted by IC ascendingly

Computer science field	IC	OOC
Scientific computing	0.168	0.678
Machine learning & pattern recognition	0.216	0.401
Natural language & speech	0.229	0.406
Bioinformatics & computational biology	0.234	0.472
Graphics	0.264	0.403
Artificial intelligence	0.281	0.420
Computer education	0.281	0.448
Multimedia	0.284	0.448
Computer vision	0.294	0.423
Security & privacy	0.308	0.435
World wide web	0.343	0.431
Human–computer interaction	0.363	0.482
Real-time & embedded systems	0.377	0.501
Information retrieval	0.393	0.455
Programming languages	0.433	0.513
Software engineering	0.434	0.501
Algorithms & theory	0.443	0.492
Networks & communications	0.450	0.502
Databases	0.456	0.549
Simulation	0.459	0.513
Distributed & parallel computing	0.466	0.540
Data mining	0.468	0.540
Operating systems	0.519	0.536
Hardware & architecture	0.579	0.610

being usually inspired and applied outside the community. The IC scores also yield many interesting insights. For example, the *Data Mining* community has many different and independent problems, and thus community communication is less converged, while its sister field *Natural Language & Speech* community (N.L.S.) is very connected and focused on languages.

Second, we analyze the distance between communities in order to understand inter-field communication, and the OOC homogeneity score. For each community, we visualize its top three citing buddies (see Fig. 1a). The visualization indeed provides many insights. For example, we can see that N.L.S. people frequently communicate with *Bio-Informatics* colleagues, while favorite buddies of *Human–Computer Interface* are in *Computer Graphics* and N.L.S. Each community provides insights of this kind.

In addition, we compare regular citations and impactful citations. By "impactful" we mean pivotal works selected as best papers in conferences (see Fig. 1b). In N.L.S., to our surprise, the impactful citations were to *Machine Learning*, *World Wide Web*, and *Programming Languages*. This could suggest "impactful" research trends to follow. Another interesting impactful citation is from

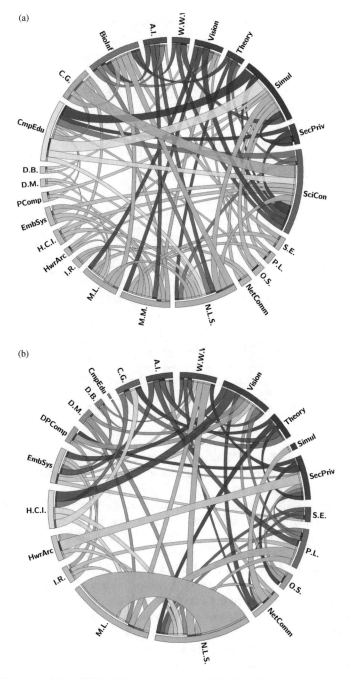

Fig. 1 Citation analysis in DBLP. (**a**) Regular citation in 24 fields in DBLP. (**b**) Best-paper citation in 24 fields in DBLP

Table 6 Connectors in the data mining community

From data mining	Author	Score
Algorithms & theory	David A. Cieslak	0.363
Artificial intelligence	D. Sculley	0.244
Bioinformatics & computational biology	Muneaki Ohshima	0.318
Computer education	Guilong Liu	0.295
Computer vision	Jianhui Chen	0.258
Databases	Xingzhi Sun	0.528
Distributed & parallel computing	David A. Cieslak	0.496
Graphics	Jianhui Chen	0.274
Hardware & architecture	Yanbo J. Wang	0.532
Human–computer interaction	Shunkai Fu	0.345
Information retrieval	Ronen Feldman	0.363
Machine learning & pattern recognition	Shunkai Fu	0.245
Multimedia	David A. Cieslak	0.289
Natural language & Speech	Ata Kaban	0.259
Networks & communications	David A. Cieslak	0.363
Operating systems	Yanbo J. Wang	0.629
Programming languages	Yanbo J. Wang	0.486
Real-Time & embedded systems	David A. Cieslak	0.387
Scientific computing	David A. Cieslak	0.321
Security & privacy	David A. Cieslak	0.301
Simulation	David A. Cieslak	0.444
Software engineering	David A. Cieslak	0.362
World wide web	Makoto Haraguchi	0.355

Hardware & Architecture to *Security & Privacy*, suggesting increasing interest in security. Also, *Vision* research might consider greater focus on *Human–Computer Interaction* for real-life impact.

Finally, we present some examples for community connectors in Table 6, for the *Data Mining* community. These authors could be considered outliers to the extent that they work between the two communities, or that their work is related to multiple research fields. For example, *David A. Cieslak* is a central researcher in *Data Mining* but has interests in a number of other fields, with researchers in different disciplines. Another example is *Jianhui Chen*. He is an active researcher in *Data Mining* with 30 publications (6 from KDD). At the same time, he is also working on dimensionality reduction and structural regularization, with papers in CVPR 2007, ICML 2009, and NIPS 2011. These works are commonly cited by the *Computer Vision* community. Interestingly, *Shunkai Fu* is connector for both *Human–Computer Interaction* and *Machine Learning & Pattern Recognition*. *Shunkai Fu* is also a founder of many mobile startups. In fact, these community connectors share similar motivations with impactful works above—trying to connect different research disciplines.

7 Conclusion

This paper reports on our development of *node embeddings*, an adaptation of representation learning, for use with nodes in social networks. Representation learning methods have been the source of impressive results in a variety of fields, but not social network analysis to our knowledge. We set out to investigate the successfulness of this approach here.

Essentially node embeddings represent a network in a d-dimensional vector space, in which each node is located independently. Links in networks are re-adjusted via our proposed EBCD using proximal measures between clusters of link's nodes. Gains recorded in community detection tasks on the resulting network indicate that our strategy can effectively re-adjust links using node embeddings in large network for more informative analysis and mining performance.

In addition, this paper also reports on large-scale experiments with a dataset covering the computer science literature. We implemented node embeddings for the DBLP citation network prior to September 2013, a network with over 2 million papers and over 2 million citations on about 300,000 papers, using 200-dimensional vectors in the representation of each node. The experimental results proved that node embedding is definitely useful, not only in social network analysis but also in general network analysis and mining.

References

1. Bengio S, Heigold G. Word embeddings for speech recognition. In: Proceedings of the 15th conference of the international speech communication association, Interspeech; 2014.
2. Blondel VD, Guillaume JL, Lambiotte R, Lefebvre E. Fast unfolding of community hierarchies in large networks. CoRR. 2008;abs/0803.0476.
3. Chen W, Zhang M, Zhang Y. Distributed feature representations for dependency parsing. IEEE Trans Audio Speech Lang Process. 2015;23(3):451–60.
4. Clauset A, Shalizi CR, Newman MEJ. Power-law distributions in empirical data. SIAM Rev. 2009;51(4):661–703.
5. Collobert R, Weston J, Bottou L, Karlen M, Kavukcuoglu K, Kuksa P. Natural language processing (almost) from scratch. J Mach Learn Res. 2011;12:2493–2537.
6. Feingold E, Good P. Encode pilot project; 2003. http://www.genome.gov/26525202.
7. Fortunato S, Lancichinetti A. Community detection algorithms: a comparative analysis: invited presentation, extended abstract. In: Proceedings of the fourth international ICST conference on performance evaluation methodologies and tools, VALUETOOLS '09. Brussels: ICST (Institute for Computer Sciences, Social-Informatics and Telecommunications Engineering); 2009. p. 27:1–2.
8. Gehrke P. The ethics and politics of speech: communication and rhetoric in the twentieth century. Carbondale: Southern Illinois University Press; 2009.
9. Goyal A, Bonchi F, Lakshmanan LVS. Approximation analysis of influence spread in social networks. CoRR. 2010;abs/1008.2005.
10. Han J. Data mining: concepts and techniques. San Francisco, CA: Morgan Kaufmann; 2005.

11. Hannun AY, Case C, Casper J, Catanzaro BC, Diamos G, Elsen E, Prenger R, Satheesh S, Sengupta S, Coates A, Ng AY. Deep speech: scaling up end-to-end speech recognition. CoRR. 2014;abs/1412.5567.

12. Hinton GE, Osindero S, Teh YW. A fast learning algorithm for deep belief nets. Neural Comput. 2006;18(7):1527–54

13. Levy O, Goldberg Y. Dependency-based word embeddings. Baltimore, MD: Association for Computational Linguistics; 2014.

14. Levy O, Goldberg Y. Neural word embedding as implicit matrix factorization. In: Ghahramani Z, Welling M, Cortes C, Lawrence N, Weinberger K, editors. Advances in neural information processing systems, vol. 27. Red Hook, NY: Curran Associates; 2014, p. 2177–85.

15. Li L, Su H, Lim Y, Li F. Object bank: an object-level image representation for high-level visual recognition. Int J Comput Vis 2014;107(1):20–39.

16. Mikolov T, Chen K, Corrado G. Dean J. Efficient estimation of word representations in vector space. CoRR. 2013;abs/1301.3781.

17. Newman MEJ. Analysis of weighted networks. Phys. Rev. E 2004;70:056131

18. Newman MEJ. Fast algorithm for detecting community structure in networks. Phys. Rev. E 2004;69:066133.

19. Newman MEJ. Finding community structure in networks using the eigenvectors of matrices. Phys. Rev. E 2006;74:036104

20. Newman MEJ, Girvan M. Finding and evaluating community structure in networks. Phys. Rev. E 2004;69:026113.

21. Orman GK, Labatut V, Cherifi H. On accuracy of community structure discovery algorithms. CoRR. 2011;abs/1112.4134.

22. Pons P, Latapy M. Computing communities in large networks using random walks (long version). In: Computer and Information Sciences-ISCIS; 2005. p. 284–93. ArXiv:arXiv:physics/0512106v1.

23. Riondato M, Kornaropoulos EM. Fast approximation of betweenness centrality through sampling. In: WSDM '14; 2014.

24. Rosvall M, Bergstrom CT. Maps of random walks on complex networks reveal community structure. Proc Natl Acad Sci USA. 2008;2007:1118.

25. Ruan, Y., Fuhry, D., Parthasarathy, S.: Efficient community detection in large networks using content and links. CoRR. 2012;abs/1212.0146.

26. Socher R, Perelygin A, Wu J, Chuang J, Manning CD, Ng AY, Potts C. Recursive deep models for semantic compositionality over a sentiment treebank. In: Proceedings of the 2013 conference on empirical methods in natural language processing. Stroudsburg, PA: Association for Computational Linguistics; 2013. p. 1631–42.

27. Sun Y, Han J. Mining heterogeneous information networks: principles and methodologies. San Rafael, CA: Morgan & Claypool; 2012.

28. Sun Y, Yu Y, Han J. Ranking-based clustering of heterogeneous information networks with star network schema. In: Proceedings of the 15th ACM SIGKDD international conference on knowledge discovery and data mining, KDD '09; 2009. p. 797–806.

29. Tian Y, Hankins RA, Patel JM. Efficient aggregation for graph summarization. In: Proceedings of the 2008 ACM SIGMOD international conference on management of data, SIGMOD '08; 2008. p. 567–80.

30. Vu T, Parker DS. Node embeddings in social network analysis. In: Proceedings of the 2015 IEEE/ACM international conference on advances in social networks analysis and mining 2015, ASONAM '15; 2015. p. 326–9.

31. Yang T, Jin R, Chi Y, Zhu S. Combining link and content for community detection: a discriminative approach. In: Proceedings of the 15th ACM SIGKDD international conference on knowledge discovery and data mining, KDD '09; 2009. p. 927–36.

32. Zhou Y, Cheng H, Yu JX. Graph clustering based on structural/attribute similarities. Proc VLDB Endow. 2009;2(1):718–729.

A LexDFS-Based Approach on Finding Compact Communities

Jean Creusefond, Thomas Largillier, and Sylvain Peyronnet

1 Introduction

Every person that has friends, acquaintances or any social ties is part of a social network. This network is used to interact with people in many different and complementary ways. The more the technology has developed, the more these interactions have begun to leave traces, creating virtual social networks, projections of the underlying social network. These networks exist in many forms: web social network, collaboration, communication, etc.

Many properties of such networks have been discovered in the last decades. The property that will be discussed in this article is that people naturally form groups, creating dense structures called communities. These communities play a central part in the structure and dynamics of the network. Indeed, we naturally classify our relationships depending on the community they belong to (family, college friends, etc.), which implies breaking the network down into small and meaningful pieces. Such structures arise in most of the naturally formed networks. Two problems arise when trying to find these communities: defining them, and designing algorithms to find them.

Strong and weak communities are objects that were defined by Radicchi et al. [31], and these definitions are commonly accepted. A strong community is defined by a group of users that have more links inside of the community than outside. The weak community definition relaxes the strong one by stating that there are more

J. Creusefond (✉) • T. Largillier
Normandy University, Caen, France
e-mail: jean.creusefond@unicaen.fr; thomas.largillier@unicaen.fr

S. Peyronnet
ix-labs, Rouen and Qwant, Paris, France
e-mail: sylvain@ix-labs.org

© Springer International Publishing AG 2017
M. Kaya et al. (eds.), *From Social Data Mining and Analysis to Prediction and Community Detection*, Lecture Notes in Social Networks,
DOI 10.1007/978-3-319-51367-6_7

links connecting the inside of a community than connecting the outside. However, the direct use of these definitions is quickly intractable, since the number of subsets of vertices following these definitions is often exponentially large. The solution is to give an ordering of these communities based on the desirable properties they feature, usually using quality functions that quantify the compliance to the property.

This paper proposes a new quality function, compactness, that measures how compact communities are, i.e. there is a small distance between individuals in the community. This property is related to communication properties: in a tightly knit group of people, important information spreads quickly to all members due to the structure of the network. This quality function can be used in situations where the effectiveness of the community structure with respect to communication spreading is important, such as marketing or political activity network analysis.

We also propose a new community detection method, based on the LexDFS traversal algorithm, that has the property of visiting clustered structures in a short time-lapse. This method is designed for social networks, but might also be applicable for other graph clustering purposes, such as biological or transport networks where dense sub-structures may appear.

The main contributions of this article are:

1. an efficient clustering algorithm that is based on the LexDFS graph traversal (Sect. 4)
2. a quality function that rates compact structures highly (Sect. 5)
3. experiments showing the practical difference between the techniques previously presented and the standard ones (Sect. 6).

This paper is an extended version of a previous article [13], featuring new proofs for axiom compliance, reworked experiments and a better comparison with existing methods, leading to new conclusions. One of the conclusions that has changed is that the LexDFS-based method is not consistently the best one with respect to compactness, but that each of these two contributions has its own merits.

2 Related Work

Some community detection algorithms have now become standard. The Girvan and Newman algorithm [18] is the first widely recognised algorithm that solved the community detection problem. *They also* introduced 2 years later [29] an algorithm optimising greedily a measure called modularity. This algorithm was then adapted for low density graphs by Clauset et al. [10]. We also note that many clustering algorithms may be used for graph community detection, such as [21]. For further reading, Fortunato [16] made a very thorough summary of the state of community detection in research.

Our community detection method is similar in spirit to random walk-based algorithms such as MCL [38] or the Pons&Latapy "walktrap" algorithm [30], where one searches for areas of the graph where random walks get "trapped". The main difference with our work is that the walks that are studied are not random, but have a memory (and favour nodes that have already been encountered in a previous neighbourhood).

One of the main foci of this paper is the property that communities are compact. This property has already been suggested as important for community detection. In [26], Leskovec et al. suggest that a low diameter for a large cluster is an indication of tightly connected nodes, a community. They compare two algorithms, and find out that the one giving lower diameter communities does not prevail when comparing conductance.

Some theoretical background exists for a diameter-based separation of the graph. For instance, Hansen and Jaumard [19] study the problem of cutting the network into two clusters while minimising the sum of their diameters. They design an algorithm that performs in a cubic complexity on a complete graph. This approach is thus not applicable to the case of real-life communities due to its high time complexity and the scaling issues that are likely to occur.

The CFinder program developed by Adamcsek et al. [1], one of the most well-known overlapping community detection software, uses another definition of a community. They consider that a community is a set of adjacent k-cliques. We note that, as with compactness, this definition does not take into account the neighbourhood of the community and define it as having a particular internal structure. However, they don't define a quality function (a set of nodes is a community or is not) and the time complexity of the algorithm is not satisfactory due to the need of finding all k-cliques of a graph.

In a similar manner, the k-core decomposition, introduced by Seidman [34] is close to the philosophy of what we are proposing. A k-core is a set of vertices in which each vertex has at least k neighbours. Decomposing the network in k-cores extracts subsets with guaranteed minimum connectivity, which can be considered as the core of communities. Our approach is to consider the core as a well-organised structure, in which communication is efficient (i.e. paths are short).

3 Notations

We use undirected graphs $G = (V, E)$, composed of a set of nodes V and edges $E \subseteq V \times V$, with no self-loop. For the sake of brevity, we call the number of nodes $n = |V|$ and the number of edges $m = |E|$. The number of edges incident to a node v is $k_v = |\{\{u, v\} \in E\}|$. We assume that nodes and edges can have a bounded number of attributes, and we note $x.a$ the way to access the attribute a of the vertex or edge x. We call satellite node a vertex $v \in V$ such that $k_v = 1$ and we call connection node a vertex which removal disconnects the graph.

A path in a graph is a sequence of connected vertices. We note the set of paths between two nodes u and v as paths$(u \in V, v \in V)$, the length of a path being defined as its size minus one (the number of edges encountered). The distance between two nodes is the minimum length of the paths between those two nodes (noted dist(u, v)), and the diameter is the maximum distance between two nodes of the graph (noted diam(G)).

A clustering C is a set of clusters (sets of nodes), and is a partition set of V. The volume of a cluster $c \in C$ is defined as Vol$(c) = \sum_{i \in C} k_i$. We define internal edges as $E(c) = (u, v) \in c^2; (u, v) \in E$. The paths are defined on clusters by using the subgraph $G_c = (c, E(c))$ that is induced by the c. Distance dist(u, v, c) and diameter diam(c) are equivalently defined.

4   LexDFS-Based Clustering Algorithm

In this section, we present our clustering method based on the LexDFS algorithm.

4.1   The Method

This section presents our efficient (sub-quadratic) clustering method that has the property of detecting the core of the communities. It is based on the LexDFS (Lexicographical Depth-First Search) algorithm introduced by Corneil and Krueger [12], a variation of the standard DFS algorithm. The difference lies in the choice of the next node to visit at each step. In the standard DFS, a node is chosen uniformly at random among the neighbours of the current node. If all the neighbours of the current node have already been visited, the neighbours of the previous node are considered, and so on.

The LexDFs algorithm makes less non-deterministic choices. Each node starts with a blank label. When visiting the (chronologically) ith node, the label "i" is added at the start of the label of all its neighbours that have not been visited. The node with the label that has the higher lexicographical order is chosen to be visited next. The related pseudo-code is presented in Algorithm 1. It uses two attributes on the nodes. The *lex* attribute is a vector of labels used to determine the priority of neighbours. The *visited* attribute marks the iteration at which the node has been visited (if it has not, it will be zero).

The LexDFS algorithm has not been studied much in the literature, except recently for theoretical research [11] to certify co-comparability orderings.

We propose to use LexDFS as a basis for a clustering method because the ordering induced by graph search features the interesting property that once in a highly connected part of the graph, the traversal often stays inside it. For clarity, we refer to the LexDFS algorithm as "algorithm", while our algorithm is referred to as "method". Because of a large number of inter-communities edges, choosing a neighbour randomly will often lead to another vertex in this highly connected part. This property is shared with the standard DFS and is closely related to the random walk properties of the communities. However, LexDFS strengthens it by the fact

```
 1: procedure LEXDFS(G, start)
Require: Graph G, Starting node start
 2:    for v ∈ V do                                          ▷Inits the attributes for every node
 3:        v.lex = ()
 4:        v.visited = 0
 5:    end for
 6:    stack = ∅
 7:    push(stack, start)
 8:    i=1
 9:    while notEmpty(stack) do
10:        node = pop(stack)
11:        node.visited = i                                         ▷Marks the node as visited
12:        array = ∅
13:        for v ∈ neighbours(node) do
14:            if v.visited = 0 then
15:                remove(stack, v)
16:                v.lex = (i, v.lex)                                      ▷Appends i to the label
17:                push(array, v)
18:            end if
19:        end for
20:        sort(array)      ▷Sorts by lexicographical order of the labels. Randomises the choice
        between equal labels
21:        push(stack, array)                                  ▷Sets the nodes of the array on top
22:        i = i+1
23:    end while
24: end procedure
```

Algorithm 1: LexDFS

that discovering a node increases the lexicographical order of its neighbours, which are mostly inside the same community. They are therefore likely to be among the next vertices to be visited.

These properties imply that nodes inside of a community are visited consecutively. A score may be computed for each edge, measuring the absolute difference of the visit of the connected nodes. We simply take the absolute value of the difference between the visit iteration, and normalise it so that a high score means two nodes are very close together, and so that the score is between 0 and 1.

$$\forall e = (u, v) \in E, \text{score}(e) = 1 - \frac{|u.\text{visited} - v.\text{visited}|}{m} \tag{1}$$

We take the mean value of this score over a few runs. Experiments show that this mean score, after a few ($\sim$ 10) runs of the LexDFS offers good topological information: filtering out the lower score edges unravels the community structure (e.g., Fig. 1). The correlation between the number of LexDFS runs and the relevance of the result is experimentally studied in Sect. 6. This property is then used in the agglomerative (bottom-up) hierarchical Algorithm 2.

Note that there are two random choices in the method: the starting node and the choice of the next node between equal labels. Having a deterministic choice at these

(a) 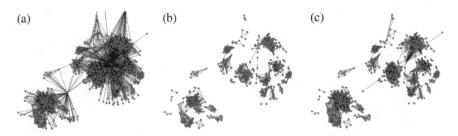 (b)   (c)

Fig. 1 State of the clusters at different iterations. (**a**) Full graph excerpt. (**b**) 1800th step. (**c**) 2700th step

```
for i ∈ [1..l] do
    LexDFS(G, randnode(G))                          ▷ Starts a LexDFS on a random node
    for (u, v) ∈ E do                               ▷ Updates the mean value of the scores
        s = 1 - |u.visited-v.visited|/m
        e.score = (e.score*(i-1)+s)/i
    end for
end for
orderedSet = E
sort(orderedSet)                                    ▷ Sorts the edges by decreasing score
C = V
while |orderedSet| > 1 do
    edge = (v₁, v₂) = pop(orderedSet)               ▷ Gets the current top-score edge
    c₁ = cluster(v₁)                                ▷ Gets the cluster of v₁
    c₂ = cluster(v₂)
    if c₁ ≠ c₂ then
        merge(c₁, c₂)
    end if
end while
```

Algorithm 2: Hierarchical clustering

decision points would induce a bias. For instance, choosing constantly a satellite node as starting point would create a large score on its edge, while they should have a low score in practice. In the same way, choosing constantly the same edge after visiting a node would induce artificial high scores for these edges.

A run of this method with ten LexDFS iterations is presented Fig. 1. The clusters are the connected components (edges are shown if and only if they are inside a cluster) and singleton clusters are hidden. The spatialisation used is the algorithm presented by Hu in [20]. The graph is an excerpt of the Facebook ego network presented in Sect. 6. We observe in Fig. 1b that a local community structure appears. Interestingly, communities span and grow separately until they connect to each other. Figure 1c shows that the two closely connected communities at the bottom left merge into one.

4.2   Complexity Analysis

The method's time complexity is in $O(m \times \log(m) + (m + n) \times l)$, where l is the number runs of the LexDFS.

4.2.1   LexDFS

LexDFS visits each node once and each edge twice. Visiting an edge (l.15–17 of Algorithm 1) is in $O(1)$ with the appropriate data structure. Indeed, if the vertex links to its position in the stack, the removal can be done in constant time. Adding a label to a vertex should also be linear if a linked list is used to represent labels. We consider that the array of neighbours is of size $n - 1$, which is the maximum number of neighbours. A variable keeps track on the actual size of the array. This data structure enables constant time addition of a vertex and standard sorting.

However, the complexity of the sorting operation (l.20) is not immediate, since the comparison is not in $O(1)$, but depends on the size of the label of the elements. We examine the case of a vertex which has d neighbours. We note that the ith neighbour in the size d array has a degree k_i and, in the worst-case scenario, it has already been discovered by all its neighbours. Therefore, the label has k_i elements and comparing the label of the ith and jth node takes $O(\min(k_i, k_j))$.

In a standard sorting algorithm such as quick sort or merge sort, two different elements are never compared twice. Taking m such as $k_m = \max_i(k_i)$, we therefore have an upper bound of the cost of the comparisons (with $\bar{k}$ being the mean degree of the neighbours):

$$\text{cost} < \sum_{i=1}^{d} \sum_{j=i+1}^{d} \min(k_i, k_j) < d \times \sum_{i=1}^{d} \min(k_i, k_m) = d^2 \times \bar{k} \qquad (2)$$

This cost is summed over all the vertices, the total cost of the comparisons is thus in $O(\sum_{v \in V} k_v^2 \bar{k}_v)$. We assume that the distribution of the degree of a neighbour of a uniformly selected vertex is seemingly the same as the uniform degree distribution, therefore $O(\bar{k}) = O(d)$. Since $\forall (a, b) \in \mathbb{R}^{+2}, a^2 + b^2 \leq (a + b)^2$, the total cost of lexicographic comparisons is in $O(n \times \bar{d}^3)$, where $\bar{d}$ is the mean degree.

The complexity of the LexDFS algorithm is therefore in $O(m + n \times \bar{d}^3)$. In social networks, the degree distribution is considered scale-free: its probability distribution is not affected by the size of the network. Therefore the mean degree may be considered constant, i.e. $O(d) = O(1)$. The final complexity of the LexDFS algorithm is thus in $O(n + m)$.

4.2.2 Cluster Merging

Computing the score of the edges is in $O(m)$. Sorting the edges given their score is in $O(m \times \log(m))$.

The successive merges of the clusters is a case of uniting disjoint sets. This problem has been solved by a quasi-linear algorithm by Tarjan in [36], in $O(m \times \alpha(m))$. Since $\alpha(m)$ is the inverse of the Ackerman function, it grows *very* slowly. Since it grows slower than a logarithmic function, the complexity of the merging is dominated by $O(m \times \log(m))$.

The complexity of the whole method is therefore in $O(m \times \log(m) + (m + n) \times l)$.

5 Quality Functions

A quality function $f : \mathscr{C} \to \mathbb{R}$ (where $\mathscr{C}$ is the set of all possible clusterings) may be defined to set a value evaluating clusterings. Its more immediate application is the comparison of the result of clustering algorithms. For a hierarchical clustering algorithm, a quality function is even more crucial: since each step produces a distinct clustering, the quality function gives us indications on which ones are the best. In this case, the quality function helps selecting relevant solutions.

5.1 Conductance

Kannan et al. [21] tried to find cuts that were meaningful for clustering. Instead of cutting the minimum number of edges, their cuts were minimal with respect to a quality function. They defined the *conductance* of a cluster, corresponding to the probability that a random walk will exit the cluster. Note that if this probability is low, the quality is high. In practice, it measures the external degree of the cluster over the volume of the cluster.

$$\phi(c \in C) = \frac{|\{(u, v) \in E, u \in c \text{ and } v \notin c\}|}{\min(\text{Vol}(c), 2m - \text{Vol}(c))} \qquad (3)$$

When considering conductance at graph-level, we normalize it in two ways. First, we use the inverse conductance, $1 - \phi(c)$, for coherence: high inverse conductance means high quality. Then, we agglomerate the scores of the communities by taking the weighted average:

$$\phi(C) = \sum_{c \in C}(1 - \phi(c))\frac{|c|}{n} \qquad (4)$$

5.2 Clustering Coefficient

The clustering coefficient was introduced by Watts and Strogatz [40] to illustrate the observation that social networks were more clustered than random networks. It corresponds, in its local version, to the average fraction of connected neighbours, *e.g.* adjacent vertices. To qualify the quality of communities, we add the restriction that the neighbours must share the same cluster as the considered node. A high clustering coefficient therefore means a highly connected cluster.

$$\text{clust}(c \in C) = avg_{v \in c} \frac{|\{u, w \in C^2, \{v, u\} \in E, \{v, w\} \in E \text{ and } \{u, w\} \in E\}|}{|\{u, w \in C^2, \{v, u\} \in E \text{ and } \{v, w\}|} \tag{5}$$

When the need for a global clustering coefficient arises, we take the weighted average as for conductance. However, since a high clustering coefficient means a high quality, there is no need for another normalisation.

5.3 Modularity

The most popular quality function in the last decade [6], [10], [17] is the *modularity*, Q defined by Newman in [29] as:

$$Q(C) = \sum_{c \in C} \left[\frac{E(c)}{m} - \left(\frac{Vol(c)}{2m} \right)^2 \right] \tag{6}$$

where $E(c)$ is the number of edges connecting the vertices of the nodes inside the cluster c.

The first part of the sum is called *coverage*. It represents the fraction of edges inside the clusters. The second part is, for each cluster, the expected value of the coverage when applying the configuration model to the full graph. It is a simple model that, given a degree distribution, connects every half edge (or "stub") to another with uniform probability. Modularity therefore detects if a group of nodes is unexpectedly tied together, but does not assume any internal structure. This perception of a community agrees with the standard definition presented in Sect. 1.

5.4 Compactness

We define here our quality function, the compactness. Modularity gives importance in a community to the internal number of edges compared to the external one. Even disregarding the issue of modularity with scaling [17], we believe that it misses an important point: the shape of the community.

Fig. 2 A simple case where
the standard definition of a
community is not satisfactory

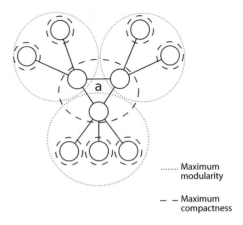

.... Maximum
        modularity

— — Maximum
      compactness

In Fig. 2, the central 3-clique a is not a community in the weak or the strong sense. Nevertheless, a human would identify the central 3-clique as a single entity, and would therefore not agree with the standard definition of a community. This example works exactly the same if the satellite nodes are any other kind of subgraph, other 3-cliques for instance. Even more convincing is the generalisation to a central n-clique, where each node is connected to n different satellites.

What makes communities in Fig. 2 stand out is their internal structure. Since every node is connected to another, communication between individuals is instantaneous. But, in a general case, information with general purpose passes through intermediates to reach its destination. We model such a transfer of information as a perfect broadcast communication process, that is the information reaches the individual i at a time t iff the information reached one of the neighbours of i at time $t - 1$. The efficiency of a connected subgraph regarding such an information transfer model can be quantified as the average transfer rate before the stable state. A lot of people reached in a little time implies a good quality of the subgraph, and a clique is the best subgraph in that case. This intuition behind it is the following: a characteristic of a group of friends or a family is that important news reaches everyone very quickly.

However, this definition makes the very strong assumption on the underlying transfer model that the communication is perfect. It is not true in most cases of observed communications, but a strong correlation between the number of neighbours that have transferred the message and the probability of transfer is natural. We therefore take into account the density of the subgraph, and we use the average edge rate for the process instead of the average node rate to quantify quality. We define a *compact community* as subsets of vertices within which vertex–vertex connections are dense, but the length of paths is small. The length of paths may be represented by the mean eccentricity (the expected average edge rate for a random source) or the diameter (the edge rate for the worse source). On this basis, we define an alternate quality measure, the *compactness* (L).

$$L(C) = \sum_{c \in C} L(c)$$

$$L(c) = \begin{cases} 0 & \text{if } E(c) = 0 \text{ or } c \text{ disconnected} \\ \dfrac{E(c)}{\text{diam}(c)} & \text{otherwise} \end{cases}$$

where diam(c) is the diameter of the sub-graph induced by the cluster c. The compactness of a cluster that has no edge (and therefore a zero diameter) is zero as well. Note that the measure can be simply normalised with a division by the total number of edges.

This quality function does not always favour the communities as defined in the introduction, weak or strong. The (normalised) measure is also close to the coverage, with the added property that the internal organisation of the cluster is taken into account. A disorganised, spread cluster has a low quality, while a compact one is considered as high quality.

The maximum compactness of a graph with n vertices is attained by an n-clique, and therefore the better clustering of an n-clique consists of a unique cluster containing the whole graph. On the other hand, when a community structure is visible, experiments showed that the whole graph is not the optimal clustering. Much higher values may be attained by regrouping the well-connected core of communities.

5.5 Comparison Between Quality Functions

We now compare compactness and modularity, showing the different choices they make in examples. We will not include conductance here, since it is very similar to modularity in all the chosen examples.

Modularity has the tendency to give a higher score to the largest possible meaningful clusters, including even satellite and connection nodes, e.g. Fig. 3. On this example, the clustering presented Fig. 3a has a better total modularity score than Fig. 3b, classifying the satellite node and the connection node.

It may be considered sound from a classification perspective of the community detection problem (every node has to be classified). However, this quality measure is not adapted when the aim is to find the connected centre of communities, and to leave the rest. The latter approach is in some sense more intuitive: when a user is only tied to one person belonging to a community, it may not be relevant to classify him in this community.

Another advantage of compactness against modularity is presented Fig. 2. Since modularity follows the standard definition of a community, unrealistic situations might arise where a clique is split between clusters. Compactness, on the other hand, detects compact clusters which is in some situations more realistic.

Fig. 3 Example: modularity (a)
(Q) and compactness (L).
(**a**) Clustering including
satellite and connection
nodes. (**b**) Clustering
excluding satellite and
connection nodes

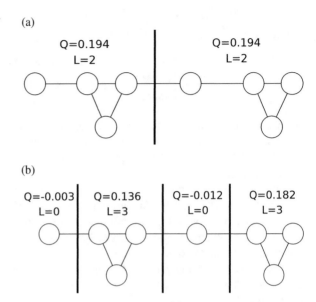

We can also compare these measures in terms of locality, that is the influence that a modification of the community (addition/deletion of an edge/node) may have on the quality score. We differentiate *local* methods, that need only to recompute values in the neighbourhood of the modification, and *global* methods that need to recompute everything. For conductance, the out-degree and the volume can be locally computed. Modularity needs the same information, it is once again a local computation. On the other hand, compactness is a global computation since even a far away node can have an important impact on the diameter. The choice to take the diameter over the mean distance implies that one node being added to a big cluster might greatly change the structure. Diameter might suddenly rise by adding satellite nodes, or decrease by adding a central node to the cluster. Therefore, a bad structure is more penalised by a bad score when taking the diameter. On the other hand, the mean distance is a more stable and a more easily approximated value.

What makes the strength of compactness is also its main drawback. Having to recompute the diameter every time a node is added induces a huge time complexity. Modularity, on the contrary, may be computed on the fly with almost no additional cost. However, it may be used only to rate a clustering algorithm and therefore be only needed for test purposes, ran on smaller instances.

5.6 Compliance to Quality Function Axioms

Van Laarhoven et al. produced a set of six axioms: **permutation**, **scale invariance**, **richness**, **monotonicity**, **locality** and **continuity** that a quality function should intuitively comply to in the context of graph clustering [39]. No unparametered quality function complying to all these axioms is presented in their work, and they proved that modularity does not comply to **locality** and **monotonicity**. We prove here that compactness complies to all of these axioms. We note that these proofs could be applied to a mean distance alternative of the compactness with little to no modification.

These axioms are defined for weighted graphs, we therefore need first to extend our definitions to weighted graphs. We comply to the notation used in [39], stating that weights are positives and a zero weight corresponds to the non-existence of an edge. The weight of the edges will correspond to a graph-dependent weight function w, $G = (V, w : V \times V \to \mathbb{R}_{\geq 0})$. Because we use a diameter computation, we need a distance function between nodes with weighted edges. This is not trivial, since the considered graph is an affinity graph (weights represent the strength of a relationship, larger strength implies closer individuals) and distances are usually defined on distance graphs (a larger strength implies a longueur route between nodes).

We therefore need to expand the underlying model. The distance function in compactness corresponds to a maximum reaching time, and in that context the strength of a relationship may be considered as a communication speed. Time being an inverse to speed, we define the path length and a distance function as follows.

$$\text{len}(v_1, \ldots, v_k) = \sum_{i=1}^{k-1} \frac{1}{w(v_i, v_{i+1})} \tag{7}$$

$$\text{dist}(u, v) = \min_{\pi \in \text{paths}(u,v)} (\text{len}(\pi)) \tag{8}$$

The numerator in the formula of compactness represents the strength of the relationships that are covered, which can be represented as a sum of the weights. The weighted version of compactness is therefore defined as follows:

$$L(C) = \sum_{c \in C} \sum_{(i,j) \in c^2} \frac{w(i,j)}{\text{diam}(c)} \tag{9}$$

The formal description of the axioms and a full proof of the compliance of compactness to all of the axioms can be found in Appendix 1. We informally describe them and explain their purpose:

Permutation invariance: The quality function is stable by homomorphism on the graph. In other words, two graphs that have the exact same structure and two clusterings that are identical with respect to the related graph will have the same score. **Intent**: A quality function should not depend on the representation of the graph. The proof that compactness complies to this axiom is presented in Appendix 1.2

Scale invariance: A quality function ranking is robust to any positive multiplicative factor applied on the weights. **Intent**: The weight of the edges often has an arbitrary scale. Any change of this scale should not impact how the quality function ranks clusterings. The proof that compactness complies to this axiom is presented in Appendix 1.3

Richness: For any clustering C of a set of node, we can create a graph on this set of nodes on which the optimal clustering is C. **Intent**: A quality function should be able to elect any kind of clustering as the optimal one. The proof that compactness complies to this axiom is presented in Appendix 1.4

Monotonicity: An increase of the weight of the edges inside clusters and a decrease of the weight of the edges between clusters leads to an increase of the quality of the clustering. **Intent**: A fundamental property of graph clusters is that they have a large internal weight and a small external weight. This axiom checks that this property is taken into account by the quality function. The proof that compactness complies to this axiom is presented in Appendix 1.5

Locality: A quality function that rates a clustering better than another on a subgraph does so independently of the rest of the graph. **Intent**: Most of the processes that naturally form clusterings are inherently local: the arrival of new nodes on one side of the graph should have an impact on far-away nodes. As underlined in [39], this axiom is related to resolution-limit properties. The article also proposes an extension of this axiom that would insure resolution-free properties, and we note that compactness satisfies this extension as well. The proof that compactness complies to this axiom is presented in Appendix 1.5

Continuity: A bounded modification of the weights of the input graph implies a bounded modification of the quality function. **Intent**: Continuity insures that the quality is robust to the noise. The proof that compactness complies to this axiom is presented in Appendix 1.5.

6 Experiments

For reproducibility, the code and the results of the experiments are available at `creusefond.users.greyc.fr`.

We used four real-world graphs for tests (from SNAP[1]): football [18] ($n = 115$, $m = 613$), Facebook [27] ($n = 4039$, $m = 88{,}234$), astro [24] ($n = 17{,}903$, $m = 196{,}972$) and enron [22] ($n = 36{,}692$, $m = 183{,}831$).

In order to keep the computation tractable, we use an approximate computation of the diameter. We select a random node in the cluster, and find the node that is the furthest away from it. The latter is considered as an eccentric point, and the highest distance stemming from it is the approximate diameter.

---

[1]http://snap.stanford.edu/data/.

6.1 Convergence

The aim of this experiment is to measure how many LexDFSs are necessary to reach convergence.

6.1.1 Description

In order to have an idea of the relationship between the convergence of the scores and the number of times a LexDFS is started, we measure at each LexDFS computation the difference of ordering of the edges. Indeed, the output ordering induces how the hierarchical clustering behaves. To that end, we apply Spearman's rank coefficient to the consecutive edge orders:

$$S(\mathbf{x}, \mathbf{y}) = 1 - \sum_{i=1}^{m} \frac{6 \times (x_i - y_i)^2}{m \times (m^2 - 1)} \tag{10}$$

where $\mathbf{x}$ and $\mathbf{y}$ are vectors containing the compared ranks.

We follow the evolution of this value over the LexDFS reiterations in order to observe the rate of convergence.

6.1.2 Results

Figure 4 presents the results of the convergence experiments. The three datasets show very similar convergence behaviour: starting from around 0.5, Spearman's coefficient rises after 3–4 iterations up to 0.95 and continues increasing to reach about 0.99 after 10 iterations. As a high convergence rate is observed, we use 10 LexDFS iterations in the other experiments.

6.2 Single Clusters Comparison

The aim of this experiment is to identify some properties of the clusters that are found by our method.

6.2.1 Description

We compare our method with a modularity optimisation algorithm, introduced by Newman et al. [29] and modified by Clauset et al. [10] for low density graphs. It is a hierarchical algorithm that starts with a community per node and then merges

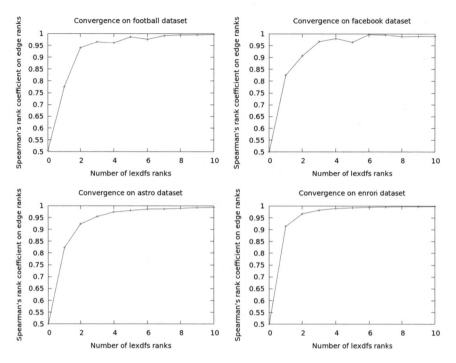

Fig. 4 Convergence of the edge ranks over multiple LexDFS

the communities that bring the best increase in modularity. The algorithm from Clauset et al. is loglinear, as ours. We observed that both implementations have similar computation time in practice.

This experiment aims to compare the quality of single clusters instead of a clustering as a whole. This method, introduced by Leskovec et al. [25] with their NCP (Network Community Profile) plots, is simply to plot the quality function related to the size of the cluster.

However, later analysis [41] showed that some quality functions, such as modularity, are less relevant when considering things at cluster-level. An explanation is that many quality functions naturally rank larger clusters higher so that the overall score weight more larger clusters. This results in difficulties to compare clusters of different sizes because their difference in quality might just be caused by the size.

To avoid this issue, and because NCP profiles already represent size as a dimension, we consider measures that are very weakly correlated with cluster size: diameter, conductance and clustering coefficient. For the latter two, we do not use the normalisation that we previously introduced.

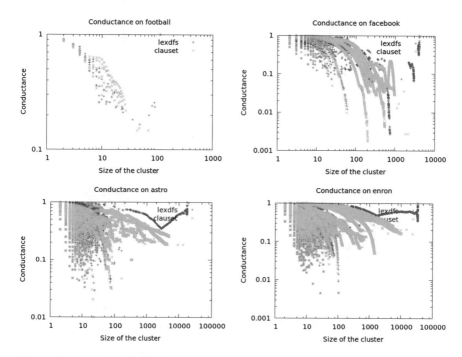

Fig. 5 Conductance for single clusters

6.2.2   Results

The conductance plot Fig. 5 shows that the modularity optimisation algorithm is better at optimising conductance. It is approximately the same in small clusters (except on the facebook dataset, where LexDFS is better), but the LexDFS-based algorithm does not perform well on large clusters regarding conductance. Both methods create small ($\sim$ 80 nodes) low conductance clusters.

The results on astro and enron show that large clusters ($>$300 nodes) with low conductance($<$0.3) are created by the modularity optimisation method. Our method does not feature such cluster, but has a similar performance on small clusters. We can also see from these plots that the LexDFS-based algorithm has the tendency to form a giant cluster quite quickly.

Figure 6 shows that small clusters generally display a wide range of clustering coefficients while larger ones are much more grouped around the global clustering coefficient. In the facebook dataset, the Clauset algorithm generates clusters with either a high ($>$ 0.5) clustering coefficient or a zero coefficient. On astro and enron, the values of both algorithms are quite similar for the lowest-size clusters. However, Clauset creates large clusters with a significantly higher clustering coefficient. We note that on the enron data, LexDFS does not create clusters of zero coefficient and

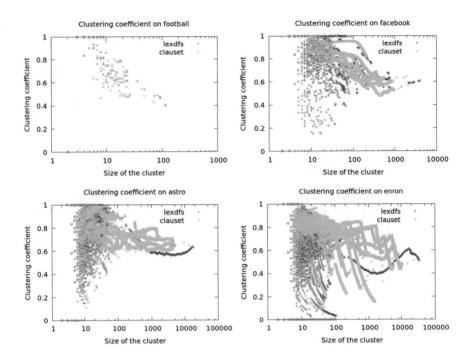

Fig. 6 Internal clustering coefficient for single clusters

of size > 25, while Clauset generates clusters that have a size of up to 100 nodes and that have near-0 clustering coefficient (since the Clauset points are over the LexDFS points, it cannot be seen on the graph).

The variety of low-size clusters in the facebook dataset is also found in the diameter measures (Fig. 7). However, middle-sized clusters in the astro and enron dataset, which tend to have a large diameter when output by the clauset algorithm, have a small diameter when generated by LexDFS.

Summary: Football is too small to differentiate LexDFS and Clauset while the astro and enron graphs feature similar characteristics. On facebook, LexDFS creates a large variety of small clusters, but they often feature relatively poor quality when compared to the output of Clauset. However, the large clusters are all very similar, which is probably due to the merging process that creates few intermediates between small clusters and a giant cluster. This small-cluster variety is also found in astro and enron, but they are very similar in quality to the small clusters of Clauset. The large clusters of Clauset have a high clustering coefficient and a low conductance, but they have a large diameter. This demonstrates that the diameter of clusters may behave very differently from other measures.

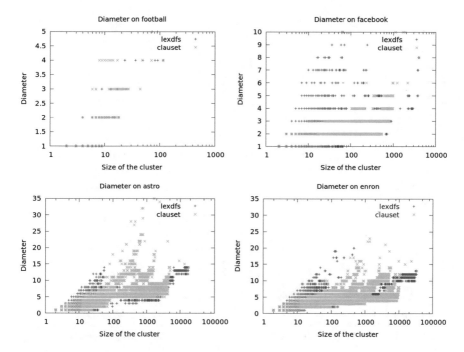

Fig. 7 Diameter for single clusters

6.3 Global Clustering Quality

The aim of this experiment is to analyse the behaviour of our method during its execution, especially regarding quality functions that may be used to select the output clustering.

6.3.1 Description

Every step of a hierarchical algorithm produces a clustering. If the application requires detecting the different communities of a network, we will be interested in the evolution of the score of a quality function during the execution of the algorithm. The global maximum score represents the best clustering according to the quality function and local variations might tell us important information about the behaviour of the algorithm. We compare the evolution of modularity, normalised clustering coefficient, normalised conductance and compactness of the algorithm over the course of the merges.

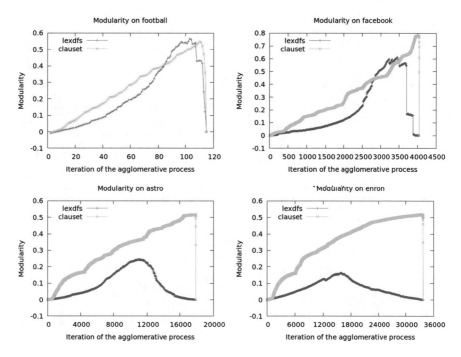

Fig. 8 Modularity over the course of the merging process

6.3.2 Results

As seen in Fig. 8, modularity has interesting properties which appear during the merging process. The figure features a "peak", an area where the quality is significantly higher than during the rest of the iterations. The fact that the global maximum is at the top of this peak means that the solution has a sense of stability: a small variation in the weighting would not lead to a completely different solution.

We can see in Fig. 9 that conductance also features peaks. However, on the astro dataset, there is a surprising peak at the end of the merging process, resulting on two "peaked" areas. The peak at the end is 1.5 times higher than the one around iteration 12,000, and is consistent across multiple runs on the same data. The cause is not clear, but it shows a possibly inconsistent behaviour of the conductance maximisation.

Figure 10 shows that for facebook and astro the internal clustering coefficient keeps rising during the merging process. Since the global maximum is at the end of the merging process, taking the maximum clustering coefficient as the final community would output the whole graph as a unique community. It is an indicator that the quality function does not suit this purpose on these graphs, since it outputs a trivial result.

Fortunately, these issues do not appear (Fig. 11). Compactness features single-peaked and non-trivial global maximum, which qualify it as an a priori good quality

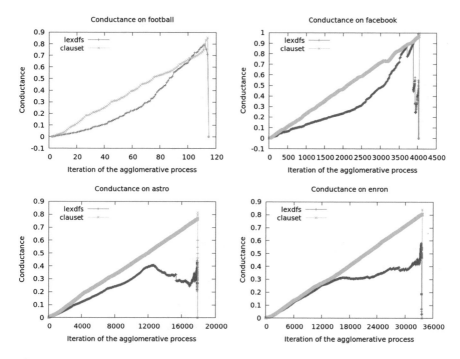

Fig. 9 Conductance over the course of the merging process

function for clustering selection. The small variations are due to the approximation used for the computation of compactness. A small error in the diameter computation may have a relatively large impact on the global compactness computation. This phenomenon was not observed on experiments where the real diameter value was computed.

We also note that the global maxima do not happen at the same place when considering different quality functions. For instance, compactness gives a global maximum at about 14,000 merges for the LexDFS algorithm, while conductance selects a late state, around 33,000 iterations. It is therefore very likely that picking a different quality function as a maximum selector will yield radically different results.

Comparing LexDFS and Clauset, we observe similar behaviours across all observed quality functions. The Clauset algorithm, in its successive states, features higher maximum modularity and conductance than our algorithm. On the other hand, our algorithm outputs clusters that are better in regards of compactness and the clustering coefficient.

We know that modularity and conductance measure internal vs. external connectivity. Compactness and the clustering coefficient, however, compute a kind of internal structure. Therefore, we conclude that our algorithm is adapted to the search of communities by their internal structure, and not by their connectivity.

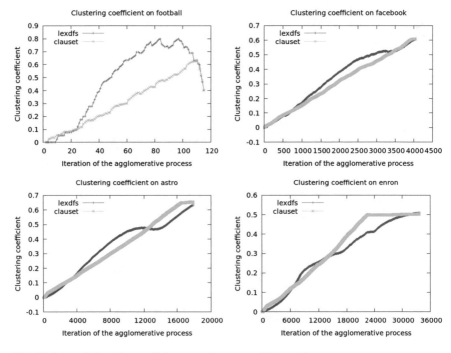

Fig. 10 Internal clustering coefficient over the course of the merging process

6.4 Performance When Compared to Ground-truth

In order to assess the performance of our algorithm in practical situations, we compare the communities found by our algorithm to ground-truth. The performance of other algorithms is computed in the same manner, to enable comparison. We also rate the performance of compactness by the similarity of its results with the ground-truth and compare it with the performance of other quality functions.

We wrote a complete paper that focuses on the methodology that we use [14]. We will provide a summary, before applying the methodology to our special case.

6.4.1 Methodology

We name overlapping clustering comparison functions, comparison methods (that is, functions that assess the resemblance of two overlapping clusterings over the same set). These methods, when used to compare a clustering and the ground-truth associated to the graph, produce a gold standard value: a quality value that is used as a reference. Their rating of the clusterings will be compared to the one given by quality functions.

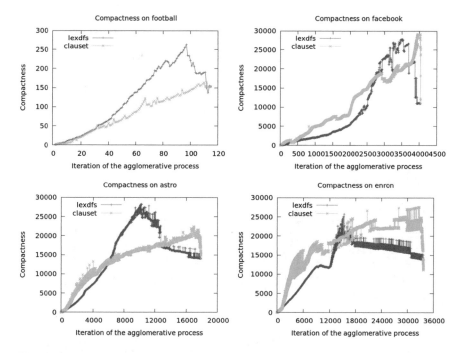

Fig. 11 Compactness over the course of the merging process

The full methodology follows:

1. Apply the community detection algorithms to the graphs
2. Compute the different qualities of the clusterings
3. Compare them to ground-truth using comparison methods, creating gold-standard values
4. Quantify the correlation between the gold standard values and the qualities

Steps 1 and 3 are used to assess the quality of a clustering. The other steps aim at comparing quality functions.

6.4.2 Components

We use ten graphs featuring ground-truth, classed by their origin (see Tables 1, 2 and 3).

We compare our algorithm with six others, including Clauset:
Louvain [5] is a widely used community detection algorithm. It finds a local maximum of modularity in a greedy fashion, by locally moving boundary vertices from one community to the other.

Table 1 Collaboration networks

Name	n	m	Nodes	Edges	Communities
DBLP[a] [41]	129,981	332,595	Authors	Co-authorships	Publication venues ⓒ
CS [8] [7]	400,657	1,428,030	Authors	Co-authorships	Publication domains ⓒ
Actors (imdb)[b] [4]	124,414	20,489,642	Actors	Co-appearances	Movies ⓒ
Github[b,c]	39,845	22,277,795	Developers	Co-contributions	Projects ⓒ

[a] http://snap.stanford.edu/data/
[b] http://konect.uni-koblenz.de
[c] https://github.com/blog/466-the-2009-github-contest

Table 2 Online social networks

Name	n	m	Nodes	Edges	Communities
LiveJournal[a] [41]	1,143,395	16,880,773	Bloggers	Following →	Explicit groups ⓒ
Youtube[a] [41]	51,204	317,393	Youtubers	Following →	Explicit groups ⓒ
Flickr [28]	368,285	11,915,549	Users	Following →	Explicit groups ⓒ

[a] http://snap.stanford.edu/data/

Table 3 Social-related networks

Name	n	m	Nodes	Edges	Communities
Amazon[a] [41]	147,510	267,135	Products	Frequent co-purchases	Categories
Football [18]	115	613	Football teams	> 1 one disputed match	Divisions
Cora[b] [35]	23,165	89.156	Scientific papers	Citations →	Categories

[a] http://snap.stanford.edu/data/
[b] http://konect.uni-koblenz.de

MCL [38] (Markov Clustering) is a matrix-based approach that operates in two steps. First, the adjacency matrix is multiplied by itself to propagate the paths: it is the expansion step. Then, a power function is applied to each element and the columns are renormalised, increasing the variance between paths and preferring short paths: it is the inflation step. By heavily relying on the low density and short paths of the graphs, this method actually runs in almost linear time, the matrix operations being quickened by a pruning.

Infomap [33] is a method based on information theory that constructs a two-level description of the network. They aim to find the first level coding that compresses the most random walks that would happen in the network. This optimisation is realised with a greedy search, refined by a simulated annealing algorithm.

3-Core [34] is an application of the core decomposition algorithm, that aims to find sets of nodes with a minimum density. A "k-core" is a set of nodes linked to the others by at least k edges, and this algorithm therefore finds 3-cores. Those cores are considered as communities.

Label propagation [32] is based on the evolution of a majority game over several time steps. Starting with one label per node, the label of each node is changed to the one featuring in the majority of its neighbours (taken at random if an equality arises). The labels of the stable states are taken as communities.

We compare compactness with nine other quality functions (including conductance, modularity and clustering coefficient):

Permanence [9] combines clustering coefficient and the maximum connectivity of a cluster towards another. **Flake-ODF** [15] is the fraction of nodes whose internal degree (the number of neighbours inside of the community) is over their external degree. **Fraction Over Median Degree (FOMD)** [41] is the fraction of nodes whose internal degree is over the total median degree. **Cut-ratio** is the external connectivity of a cluster over the same external connectivity if the graph were complete. **Surprise** [2] is the divergence between the fraction of internal edges and the probability that two nodes chosen randomly are inside the same community. **Significance** [37] is the divergence between internal density of edges and global density.

We used two comparison methods, that is functions that measure the similarity between two clusterings (C_1 and C_2).:

- The **Normalised Mutual Information (NMI)** [23] is a measure based on information theory that finds for each cluster in C_1 the cluster in C_2 that minimises mutual information. The final score is a normalised and weighted average over all clusters of the mutual information.
- The **F-BCubed (fb3)** [3] considers each vertex independently, and measures the number of its neighbours in C_1 that are also its neighbours in C_2. The final score is the average over all vertices. The overlapping version extends this concept by taking into account the number of clusters that a vertex has in common with its neighbours.

6.4.3   Evaluation of Compactness

Table 4 shows the correlation between the golden values of the F-BCubed and the different qualities, based on the output of the algorithm. We can see that compactness behaves relatively well on two graphs: football and github. There is no apparent link between the two networks: football represents the games played by football teams while github represents collaboration between users on projects. However, when analysing their structure, it appears that they both have a strong underlying bipartite structure: they are very close to unions of cliques, where the cliques are the (overlapping) communities.

But another network, actors, has the same kind of underlying structure. And, for these three graphs, two other quality functions have better or equal results (surprise and signature). All in all, it therefore seems that even if compactness is theoretically interesting, the situation in which it will be useful is yet to discover.

Table 4 Spearman's coefficient of the fb3(ground truth, algorithms) compared to the results of quality functions

file\quality	cc	fb3	mod	nmi	perm	sign	cond	FOMD	comp	cut_ratio	f-odf	sur
CS	-0.50	1.00	-0.14	0.82	0.14	-0.75	0.39	0.75	-0.61	0.59	0.18	-0.93
actors	0.29	1.00	-0.07	0.46	-0.79	0.43	-0.79	-0.79	0.18	-0.79	-0.64	0.36
amazon	-0.86	1.00	-0.04	0.03	-0.04	-0.89	0.00	0.25	-0.93	-0.01	0.00	-0.79
cora	-0.64	1.00	0.04	0.69	0.29	-0.75	0.50	0.89	-0.75	0.79	0.50	-0.75
dblp	-0.68	1.00	-0.79	0.89	-0.79	-0.57	-0.64	-0.32	-0.07	-0.67	-0.61	-0.71
flickr	0.18	1.00	-0.21	-0.71	0.07	0.04	0.07	0.46	0.39	0.60	0.07	0.29
football	1.00	1.00	0.68	0.38	0.25	1.00	-0.96	-0.05	1.00	-0.21	-0.93	1.00
github	-0.29	1.00	0.39	-0.36	-0.57	0.71	-0.61	-0.79	0.71	-0.57	-0.93	0.71
lj	0.29	1.00	-0.54	-0.86	-0.14	0.39	-0.46	-0.36	-0.11	-0.38	-0.46	0.32
youtube	0.04	1.00	-0.86	-0.89	-0.61	-0.07	-0.54	0.04	0.14	-0.19	-0.54	-0.32

source: [14]

Table 5 NMI ranks of the different maximums

	Actors(9)	Amazon	Cora	CS	dblp	Football	Flickr(9)	github	lj	Youtube
$lexdfs_{comp}$	5	7	9	10	3	3	5	4	–	7
$lexdfs_{cc}$	–	1	3	3	8	7	–	–	7	10
$lexdfs_{mod}$	1	6	7	7	2	6	4	1	4	6
$lexdfs_{cond}$	2	2	2	2	7	9	9	2	5	8

Table 6 F-BCubed ranks of the different maximums

	Actors(9)	Amazon	Cora	CS	dblp	Football	Flickr(9)	github	lj(9)	Youtube
$lexdfs_{comp}$	1	2	7	7.5	2	5	7	1	–	4
$lexdfs_{cc}$	–	5	1	1	4	10	–	–	8	10
$lexdfs_{mod}$	3	4	7	7.5	3	6	8	3	4	5
$lexdfs_{cond}$	9	3	3	4	5	8	9	9	9	9

6.4.4 Evaluation of the LexDFS Algorithm

We consider similarity between the clustering output by the different maxima and the ground-truth. This measures how realistic the clusterings output by the algorithms is.

Due to external constraints, we stopped the algorithms that were still running after 3 days. For some of the graphs, computing the number of triangles or the approximated diameters was too long.

Tables 5 and 6 present the results of the NMI and the F-BCubed, respectively, applied to the ground-truth and the output of the algorithms on the base graph. The cells contain the rank relative to the other algorithms. The runs that did not finish are marked with a "–" and therefore the associated columns have a maximal rank of 9 instead of 10.

As suspected, using maxima from different quality functions changes the ranking, sometimes radically (e.g. from 2 to 10 in cora, NMI). We can also see that

NMI and F-BCubed yield significantly different results, with an average of a 2.4 difference in the rankings. Each score that will be cited henceforth will be in the form *scorenmi/scorefbcubed*.

While it did not behave well in the global clustering experiment (see Sect. 6.3), the clustering coefficient maxima have strong results on amazon (1/5), cora (3/1) and CS (3/1). Conductance maximum also has very good scores on these graphs (2/3, 2/3 and 2/4). Both quality functions yield, however, very bad results on the rest of the graphs.

Compactness and Modularity have relatively similar results: they have good scores on actors (5/1, 1/3) and github (4/1, 1/3). They are mostly average on the rest, except on amazon where all of the lexdfs algorithms had good scores.

Summary: It appears that comparing multiple quality functions as maxima opens up to many different relevant clusterings. The clustering coefficient and the conductance maxima perform very well on CS and cora, that were already identified as close to each other [14]. The compactness and modularity maxima have a similar pattern on actors and github networks, that are similar to each other due to their strong underlying bipartite structure.

7 Conclusion

We presented a quality function that measures how compact communities are. We proved that this quality function satisfies axioms that indicate that its behaviour is intuitive. This quality function is not the most prevalent property of naturally formed labelled clusters (ground-truth). However, the fact that it represents a form of network efficiency might be relevant for communication applications.

We defined an efficient community detection algorithm. We find that it outperforms standard algorithms in some tests, notably when using ground-truths. We also showed that it produced a large variety of small clusters that often have a good internal structure.

Acknowledgements The authors thank Loïck Lhote for his help with the proof of continuity.

1 Appendix: Proofs of Axioms Compliance

For technical reasons, we also need to define compactness when applied on unconnected clusters. The quality of a cluster on which information cannot spread is the lowest in our model, therefore the quality of a disconnected cluster is defined as a zero value.

1.1 Specific Notations

Some of previous notations omitted the graph dependency for brevity (length, distance, diameter, compactness, etc.). When there is ambiguity, the graph will be specified as a subscript.

A connected graph in this context is a graph $G = (V, w)$ for which $\forall (u, v) \in V \times V$, $\text{dist}(u, v)$ is defined, or equivalently there exist a path $\pi \in \text{paths}(u, v)$ between u and v such that $\forall i \in [0 : |\pi| - 1]$, $w(a_i, a_{i+1}) > 0$.

$\mathscr{P}(V)$ is the powerset of a set V of nodes, that is the set of all possible clusters.

$\mathscr{C}(V)$ is the set of possible partitions of a set V of nodes, that is $\{\{c_1 \in \mathscr{P}(V), \ldots, c_{|c|} \mathscr{P}(V)\}, \cup c_i = V, \cap c_i = \emptyset\}$.

1.2 Permutation Invariance

Definition 1 A graph clustering quality function Q is **permutation invariant** if for all graphs $G = (V, w)$ and all isomorphisms $f : V \to V'$, it is the case that $Q_G(C) = Q_{f(G)}(f(C))$ where f is extended to graphs and clusterings by $f(C) = \{\{f(i)|i \in c\}|c \in C\}$ and $f((V, w)) = (V', (i, j) \to w(f^{-1}(i), f^{-1}(j)))$

Proposition 1 *Compactness is permutation invariant*

Sketch of proof: Compactness only uses internal edges as an input, therefore it does not depend on representation.

Proof First, distances on weighted graphs as defined previously are permutation invariant:

$$\text{dist}_{f(G)}(f(u), f(v)) = \min_{\pi \in \text{paths}_{f(G)}(u,v)} (\text{len}_{f(G)}(\pi))$$

$$= \min_{\pi \in \text{paths}_{f(G)}(u,v)} (\text{len}_G(f^{-1}(\pi)))$$

$$= \text{dist}_G(u, v)$$

Compactness is permutation invariant:

$$L_{f(G)}(f(C)) = \sum_{f(c) \in f(C)} \sum_{(f(i), f(j)) \in f(c)^2} \frac{w(f(i), f(j))}{\max_{(f(u), f(v)) \in f(C)^2} (\text{dist}_{f(G)}(f(u), f(v)))}$$

$$= \sum_{f(c) \in f(C)} \sum_{(f(i), f(j)) \in f(c)^2} \frac{w(i, j)}{max_{(f(u), f(v)) \in f(C)^2} (dist_G(u, v))}$$

$$L_{f(G)}(f(C)) = L_G(C)$$

1.3 Scale Invariance

Definition 2 A graph clustering quality function Q is **scale invariant** if for all graphs $G = (V, E)$, all clusterings C_1, C_2 of G and all constants $\alpha > 0$, $Q_G(C_1) \leq Q_G(C_2)$ if and only if $Q_{\alpha G}(C_1) \leq Q_{\alpha G}(C_2)$, where $\alpha G = (V, (i, j) \rightarrow \alpha w(i, j))$ is a graph with edge weights scaled by a factor α.

Proposition 2 *Compactness is scale invariant*

Sketch of proof: Multiplying the edges by a α implies that the diameter is multiplied by $1/\alpha$. Since the numerator of any element of the sum of compactness is multiplied by α, the total score is multiplied by α^2, therefore the order of the clusterings rank is kept.

Proof

$$\text{len}_{\alpha G}(v_1, \ldots, v_k) = \sum_{i=1}^{k-1} \frac{1}{\alpha w(v_i, v_{i+1})}$$

$$\text{len}_{\alpha G}(v_1, \ldots, v_k) = \frac{\text{len}_G(v_1, \ldots, v_k)}{\alpha}$$

Since the lengths of the paths are linearly correlated, the minimum paths are the same in G and in αG, therefore:

$$\text{dist}_{\alpha G}(u, v) = \min_{\pi \in \text{paths}_{\alpha G}(u,v)} (\text{len}_{\alpha G}(\pi))$$

$$= \min_{\pi \in \text{paths}_{\alpha G}(u,v)} \left(\frac{\text{len}_G(\pi)}{\alpha} \right)$$

$$= \frac{\text{dist}_G(u, v)}{\alpha}$$

Using the same reasoning, $\text{diam}_{\alpha G}(c) = \text{diam}_G(c)/\alpha$ the compactness can be written as:

$$L_{\alpha G}(C) = \sum_{c \in C} \sum_{(i,j) \in c^2} \frac{w_{\alpha G}(i, j)}{\text{diam}_{\alpha G}(c)}$$

$$= \sum_{c \in C} \sum_{(i,j) \in c \times c} \alpha^2 \frac{w_G(i, j)}{\text{diam}_G(c)}$$

$$= \alpha^2 L_G(c)$$

Therefore, for all clusterings C_1, C_2, if $L_G(C_1) \geq L_G(C_2)$ then $\alpha^2 L_G(C_1) \geq \alpha^2 L_G(C_2)$, implying $L_{\alpha G}(C_1) \geq L_{\alpha G}(C_2)$.

1.4   Richness

Definition 3 A graph clustering quality function Q is **rich** if for all sets V and all non-trivial partitions C^* of V, there is a graph $G = (V, w)$ such that C^* is the Q-optimal clustering of V, i.e. $\mathrm{argmax}_C Q_G(C) = C^*$

Proposition 3 *Compactness is rich*

Sketch of proof: For a graph in which all clusters in C are cliques and there is no edge between clusters, C is the most compact clustering since any other has either:

- disconnected clusters, which have a zero compactness, and separating them into connected clusters improves the compactness
- multiple clusters including nodes belonging in the same clique, and merging these clusters adds the edges that are between them to the original score

Proof $\forall C^* \in \mathscr{C}(V)$, let $G = (V, w)$ be a graph such that, $\forall (i, j) \in V^2$, $w(i, j) = 1$ if $\exists c \in C^*$ such that $i \in c$ and $j \in c$ (i and j belong to the same cluster) and $w(i, j) = 0$ otherwise. G is therefore a clique graph (all connected components are cliques), where the disconnected cliques are the clusters in C^*. Let C be an optimal clustering of G w.r.t. compactness, i.e. $\mathrm{argmax}_{D \in \mathscr{C}(V)} L_G(D) = C$.

If $\exists c \in C$ such that $\exists (i, j) \in c^2$, $w(i, j) = 0$, then i and j are not in the same cluster in C^*, and c is disconnected therefore its compactness equals to zero. When separating c into multiple connected clusters $(c_1, \ldots, c_k)$, note that compactness of any of these clusters is positive, and therefore greater or equal to the compactness of c. Calling C' the clustering such that $C' = C \backslash c \cup \{c_1, \ldots, c_k\}$. Therefore

$$L(C') = L(C \backslash c) + L(\{c_1, \ldots, c_k\})$$
$$\geq L(C \backslash c) + L(c)$$
$$\Leftrightarrow L(C') \geq L(C)$$

The scores are equals iff all nodes in c have zero degree. Therefore, $\forall c \in C$, c is connected in G or composed of zero-degree nodes (connectedness condition).

If $\exists (c_1, c_2) \in C^2$, $c_1 \neq c_2$ such that $\exists i \in c_1, j \in c_2$, $w(i, j) = 1$ (i.e. there is an edge between two clusters in C), then i and j are in the same cluster in C^* and in different clusters in C. Because the nodes i and j have a non-zero degree, the clusters c_1 and c_2 are both connected. Noting that the distance between any connected pair of nodes is always 1 on this graph, we have $\mathrm{diam}(c_1) = \mathrm{diam}(c_2) = \mathrm{diam}(c_1 \cup c_2) = 1$. We call C' the clustering corresponding to C where the c_1 and c_2 has been replaced by a fusion of them, that is $C' = C \backslash \{c_1, c_2\} \cup \{c_1 \cup c_2\}$.

$$L(C') = L(C \backslash \{c_1, c_2\}) + L(\{c_1 \cup c_2\})$$
$$\geq L(C \backslash \{c_1, c_2\}) + L(c_1) + L(c_2) + \frac{w(i, j)}{\mathrm{diam}(\{c_1 \cup c_2\})}$$

$$> L(C \backslash \{c_1, c_2\}) + L(c_1) + L(c_2)$$
$$\Leftrightarrow L(C') > L(C)$$

Since C is optimal w.r.t. L, there is no edge between the clusters of C, which is equivalent to $\forall c \in C, \exists c' \in C^*, c' \subseteq c$ (maximality condition).

Both conditions imply that any cluster in C is either a maximal connected component (and therefore a cluster in C^*) or a set of zero-degree nodes. Since any set containing a zero-degree node has the same compactness (zero), C^* has the same compactness as any other maximal clustering. Therefore, C^* is a maximum-compactness clustering of G.

1.5 Monotonicity

Definition 4 Let $G = (V, w)$ be a graph and C a clustering of G. A graph $G' = (V, w')$ is a C-**consistent improvement** of G if for all nodes i and j, $w'(i, j) \geq w(i, j)$ whenever i is in the same community as j and $w'(i, j) \leq w(i, j)$ whenever i is not the in the same community as j.

Definition 5 A graph clustering quality function Q is **monotonic** if for all graphs G, all clusterings C of G and all C-consistent improvements G' of G it is the case that $Q_{G'}(C) \geq Q_G(C)$

Proposition 4 *Compactness is monotonic*

Sketch of proof: Compactness is not influenced by between-clusters weights, and increasing the weight inside clusters can only decrease or not affect the diameter. Therefore, compactness is either unaffected or increased by such a modification.

Proof Due to the insensibility of compactness to the weight of external edges, we will focus on inter-cluster edges.

$\forall c \in C, \forall \Pi$ a path in the subgraph of G induced by C, $len_G(\Pi) \geq len_{G'}(\Pi)$ (since all weights have increased or stayed the same). Since this is true for all paths, $diam_G(c) \geq diam_{G'}(c)$. Therefore, $L_G(c) = \sum_{i,j \in c^2} w(i, j)/diam_G(c) \leq \sum_{i,j \in c^2} w'(i, j)/diam_{G'}(c) = L_{G'}(c)$.

A consistent improvement thus implies an equal or increased compactness, which proves monotonicity.

1.6 Locality

Definition 6 Let $G_1 = (V_1, w_1)$ and $G_2 = (V_2, w_2)$ be two graphs and let $V_a \subseteq V_1 \cap V_2$ be a subset of the common nodes. We say that the graphs **agree** on V_a if $w_1(i, j) = w_2(i, j)$ for all $i, j \in V_a$. We say that the graphs also **agree on the neighbourhood** of V_a if

- $w_1(i,j) = w_2(i,j)$ for all $i \in V_a$ and $j \in V_1 \cap V_2$,
- $w_1(i,j) = 0$ for all $i \in V_a$ and $j \in V_1 \backslash V_2$, and
- $w_2(i,j) = 0$ for all $i \in V_a$ and $j \in V_2 \backslash V_1$.

This means that for nodes in V_a the weights and endpoints of incident edges are exactly the same in the two graphs.

Definition 7 A graph clustering quality function Q is **local** if for all graphs $G_1 = (V_1, w_1)$ and $G_2 = (V_2, w_2)$ that agree on a set V_a and its neighbourhood, and for all clusterings C_a, D_a of V_a, C_1 of $V_1 \backslash V_a$ and C_2 of $V_2 \backslash V_a$, if $Q_{G_1}(C_a \cup C_1) \geq Q_{G_1}(D_a \cup C_1)$ then $Q_{G_2}(C_a \cup C_2) \geq Q_{G_2}(D_a \cup C_2)$.

Proposition 5 *Compactness is local*

   Sketch of proof: Thanks to additivity properties of compactness and the fact that it only uses internal data, any clustering preference on a graph G is kept on a graph G' that would include it.

Proof Let $G_1 = (V_1, w_1)$ and $G_2 = (V_2, w_2)$ be two graphs that agree on a set V_a and its neighbourhood. By definition, $\forall (i,j) \in V_a^2$, $w_1(i,j) = w_2(i,j)$. Since compactness only takes into account internal edges and internal paths, and since G_1 and G_2 agree on V_a and its internal edges, $\forall c \in \mathscr{P}(V_a)$, $L_{G_1}(c) = L_{G_2}(c)$. Therefore, $\forall C \in \mathscr{C}(V_a)$, $L_{G_1}(C) = L_{G_2}(C)$.
   We immediately get $\forall (C_a, D_a) \in \mathscr{C}(V_a)^2$, $\forall C_1 \in \mathscr{C}(V_1 \backslash V_a)$ and $\forall C_2 \in \mathscr{C}(V_2 \backslash V_a)$

$$L_{G_1}(C_a \cup C_1) \geq L_{G_1}(D_a \cup C_1)$$
$$\Leftrightarrow L_{G_1}(C_a) + L_{G_1}(C_1) \geq L_{G_1}(D_a) + L_{G_1}(C_1)$$
$$\Leftrightarrow L_{G_1}(C_a) \geq L_{G_1}(D_a)$$
$$\Leftrightarrow L_{G_2}(C_a \cup C_2) \geq L_{G_2}(D_a \cup C_2)$$

1.7 Continuity

Definition 8 A quality function Q is **continuous** if a small change in the graph leads to a small change in the quality. Formally, Q is continuous if for every $\epsilon > 0$ and every graph $G = (V, w)$, there exists a $\delta > 0$ such that for all graphs $G' = (V, w')$, if $w(i,j) - \delta < w'(i,j) < w(i,j) + \delta$ for all nodes i and j, then $Q_{G'}(C) - \epsilon < Q_G(C) < Q_{G'}(C) + \epsilon$ for all clusterings C of G.

Proposition 6 *Compactness is continuous*

Sketch of proof: First, we prove that the distance function is continuous for connected graphs. To that aim, we show that this distance on any Cauchy sequence of graphs converging to the graph converges to the distance. Diameter is therefore continuous and continuity is insured on disconnected graphs by showing that the diameter goes to infinity when any graph gets close to disconnected.

Proof We note that this definition of continuity corresponds to the standard continuity of a multivariate function, with the distance between two graphs being the maximum of the absolute difference in edge weights. We call this distance function $d(G, G')$. Therefore, we can use known properties, such as the continuity of the combination of continuous functions, etc.

Lemma 1 *For a connected graph* $G = (V, w)$, $\forall (a, b) \in V \times V$, $dist_G(a, b)$ *is continuous.*

Proof Let $G_n = (V, w_n)$ be a Cauchy sequence of graphs. Then, $\forall (i, j) \in V^2$, $(w_n(i, j))_{n \in \mathbb{N}}$ is also a Cauchy sequence, therefore $\exists w$ such that $w_n(i, j) \to w(i, j)$ (in this context, $\to$ means "converge to") and a graph $G = (V, w)$ such as $G_n \to G$. We assume G to be connected.

$\forall (a, b) \in V^2$, let $\Pi = (a_0 = a, a_1, \ldots, a_{k-1}, a_k = b)$ be a path such as $dist_G(a, b) = len_G(\Pi)$, that is a minimal path in G between a and b. If $\exists i \in [0 : k-1]$ such as $w(a_i, a_{i+1}) = 0$, then $len_G(\Pi)$ is undefined and therefore it is not a minimal path. Since a minimal path exists (due to the definition of a connected graph), $\forall i \in [0 : k-1]$, $w(a_i, a_{i+1}) > 0$.

Since $\forall i \in [0 : k-1]$, $w_n(a_i, a_{i+1}) \neq 0$, then $f(x_0, \ldots, x_{k-1}) = \sum_{i \in [0:k-1]} \frac{1}{x_i}$ is continuous in $(w(a_0, a_1), \ldots, w(a_{k-1}, a_k))$. Therefore, $len_{G_n}(\Pi) \to len_G(\Pi)$.

Since $dist_{G_n}(a, b) \leq len_{G_n}(\Pi)$,

$$\limsup_{n \to +\infty} dist_{G_n}(a, b) \leq \limsup_{n \to +\infty} len_{G_n}(\Pi)$$

$$= \lim_{n \to +\infty} len_{G_n}(\Pi) = len_G(\Pi) = dist_G(a, b)$$

$$\limsup_{n \to +\infty} dist_{G_n}(a, b) \leq dist_G(a, b)$$

$$\liminf_{n \to +\infty} dist_{G_n}(a, b) \leq \limsup_{n \to +\infty} dist_{G_n}(a, b), \text{ therefore}$$

$$\liminf_{n \to +\infty} dist_{G_n}(a, b) \leq \limsup_{n \to +\infty} dist_{G_n}(a, b) \leq dist_G(a, b) \tag{11}$$

Let $\epsilon > 0$. Since $\limsup_{n \to +\infty} dist_{G_n}(a, b) \leq dist_G(a, b)$, $\exists n_0 \in \mathbb{N}$ such that $\forall n \geq n_0$, $dist_{G_n}(a, b) \leq (1 + \epsilon) dist_G(a, b)$.

Let $n_1 \in \mathbb{N}$ such that $\forall n \geq n_1$, $d(G, G') \leq \dfrac{1}{2 \times dist_G(a, b)}$.

By definition, $\forall (i, j) \in V \times V$, $|w_n(i, j) - w(i, j)| \leq \dfrac{1}{2 \times dist_G(a, b)}$.

For $n \in \mathbb{N}$, let Π_n be a path such as $dist_{G_n}(a, b) = len_{G_n}(\Pi_n)$. If Π_n is not a path in G for $n \geq max(n_0, n_1)$, then for $\Pi_n = (a_0^{(n)}, \ldots, a_{k_n}^{(n)})$, $\exists i \in [0 : k_n - 1]$ such as $w_n(a_i^{(n)}, a_{i+1}^{(n)}) \neq 0$ and $w(a_i^{(n)}, a_{i+1}^{(n)}) = 0$. In that case,

$$|w_n(a_i^{(n)}, a_{i+1}^{(n)})| = |w_n(a_i^{(n)}, a_i^{(n)}) - w(a_i^{(n)}, a_i^{(n)})| \leq \frac{1}{2 \times \text{dist}_G(a, b)}$$

$$\Rightarrow \text{len}_{G_n}(\Pi_n) \geq \frac{1}{w_n(a_i^{(n)}, a_{i+1}^{(n)})} \geq 2 \times \text{dist}_G(a, b)$$

However, this is contradictory, since $\text{len}_{G_n}(\Pi_n) = \text{dist}_{G_n}(a, b) \leq (1 + \epsilon)\text{dist}_G(a, b)$ for $n \geq n_0$. Then, for $n \geq \max(n_0, n_1)$, Π_n is also a path in G.

Let $n_3 \in \mathbb{N}$ such that $\forall n \geq n_3$, $d(G, G_n) \leq \epsilon w_{\min}$ with $w_{\min} = \min_{(u,v) \in V^2, w(u,v) \neq 0} w(u, v)$. First, we note that

$$\forall (u, v) \in V^2, w_n(u, v) \geq w(u, v) - \epsilon w_{\min} \geq (1 - \epsilon)w_{\min}$$

Then

$$|\text{len}_{G_n}(\Pi_n) - \text{len}_G(\Pi_n)| = |\sum_{i \in [0:k_n-1]} \frac{1}{w_n(a_i^{(n)}, a_{i+1}^{(n)})} - \frac{1}{w(a_i^{(n)}, a_{i+1}^{(n)})}|$$

$$= \sum_{i \in [0:k_n-1]} \frac{|w_n(a_i^{(n)}, a_{i+1}^{(n)}) - w(a_i^{(n)}, a_{i+1}^{(n)})|}{w_n(a_i^{(n)}, a_{i+1}^{(n)}) \times w(a_i^{(n)}, a_{i+1}^{(n)})}$$

$$\leq \sum_{i \in [0:k_n-1]} \frac{\epsilon w_{\min}}{w_n(a_i^{(n)}, a_{i+1}^{(n)}) \times w(a_i^{(n)}, a_{i+1}^{(n)})}$$

$$\leq \sum_{i \in [0:k_n-1]} \frac{\epsilon w_{\min}}{(1 - \epsilon)w_{\min} \times w_{\min}}$$

$$|\text{len}_{G_n}(\Pi_n) - \text{len}_G(\Pi_n)| \leq \frac{|V|\epsilon}{w_{\min}(1 - \epsilon)}$$

Since $\text{len}_G(\Pi_n) \geq \text{dist}_G(a, b)$ (it is a path between a and b), then $\text{len}_{G_n}(\Pi_n) \geq \text{dist}_G(a, b) - \frac{|V|\epsilon}{w_{\min}(1 - \epsilon)}$. Therefore, $\liminf_{n \to +\infty} \text{len}_{G_n}(\Pi_n) \geq \text{dist}_G(a, b)$. Combined with Eq. 11:

$$\limsup_{n \to +\infty} \text{dist}_{G_n}(a, b) \leq \text{dist}_G(a, b) \leq \liminf_{n \to +\infty} \text{len}_{G_n}(\Pi_n)$$

$$\Rightarrow \lim_{n \to +\infty} \text{dist}_{G_n}(a, b) = \text{dist}_G(a, b)$$

which proves that the distance between any two nodes in a connected graph is continuous.

End of the proof of Lemma 1

We now prove continuity of the function on unconnected clusters. In order to simplify notations, we directly work on the induced subgraphs, and we extend L to take a graph as an input:

$$L(G) = \begin{cases} 0 & \text{if } |V| = 1 \text{ or } G \text{ disconnected} \\ \dfrac{\sum_{(u,v)\in V^2} w(u, v)}{\text{diam}(G)} & \text{otherwise} \end{cases}$$

From Lemma 1, we know that $\text{dist}(u, v)$ is continuous for all connected graphs, and $\text{dist}(u, v) > 0$. The maximum of continuous functions is continuous, which means that $\text{diam}(G)$ is continuous for all connected graphs and $\text{diam}(G) > 0$. The combination of continuous functions is continuous, and $1/x$ is continuous on $\mathbb{R}^+$. We conclude that $L(G)$ is continuous on all connected graphs.

We now prove that $L(G)$ is continuous on unconnected graphs. Just as in Lemma 1, we take a Cauchy sequence of graphs $(G_n)_{n\in\mathbb{N}} : G_n = (V, w_n) \to G = (V, w)$, but with G disconnected. For all $n \in \mathbb{N}$, if G_n is disconnected, $L(G_n) = 0 = L(G)$.

Since G is disconnected, $\exists(u, v) \in V \times V$, for all paths $\pi \in \text{paths}(u, v)$ between u and v, $\exists i \in [0 : k - 1]$ such that $w(a_i, a_{i+1}) = 0$. If G_n is not disconnected, there exists a minimal path $\Pi_n = (a_0 = u, a_1, \ldots, a_k = v) \in \text{paths}(u, v)$ ($\text{len}(\Pi_n) = \text{diam}(G_n)$) such that $\forall i \in [0 : k - 1]$, $w_n(a_i^{(n)}, a_{i+1}^{(n)}) > 0$. By definition, $\text{len}(\Pi_n) = \sum_{i\in[0:k-1]} \dfrac{1}{w_n(a_i^{(n)}, a_{i+1}^{(n)})} > \dfrac{1}{w_{min}^{(n)}}$ where $w_{min}^{(n)} = \min_{i\in[0:k-1]}(w_n(a_i^{(n)}, a_{i+1}^{(n)}))$.

Since G_n converges to G, and G disconnected, $\lim_{n\to+\infty} w_{min}^{(n)} = 0^+$. Therefore, $\lim_{n\to+\infty} \text{dist}_{G_n}(u, v) = \lim_{n\to+\infty} \min_{\pi\in\text{paths}(u,v)} \text{len}_{G_n}(\pi) = +\infty$.

Since the diameter is the maximum of the distances between all pairs of nodes, $\lim_{n\to+\infty} \text{diam}(G_n) = +\infty$. By definition of the Cauchy sequence, $\lim_{n\to+\infty} \sum_{(u,v)\in V^2} w_n(u, v) = \sum_{(u,v)\in V^2} w(u, v)$. Therefore,

$$\lim_{n\to+\infty} L(G_n) = \sum_{(u,v)\in V^2} w_n(u, v)/\text{diam}(G_n) = 0 = L(G)$$

which implies that for all disconnected graph G, L is continuous on G.

Since compactness is the sum of $L(G)$ applied to subgraphs induced by the clustering, compactness is continuous.

References

1. Adamcsek B, Palla G, Farkas I, Derényi I, Vicsek T. CFinder: locating cliques and overlapping modules in biological networks. Bioinformatics. 2006;22(8):1021–23
2. Aldecoa R, Marín I. Surprise maximization reveals the community structure of complex networks. Sci Rep 2013;3. http://www.nature.com/articles/srep01060?WT.ec_id=SREP-631-20130201 and http://www.nature.com/articles/srep02930
3. Bagga A, Baldwin B. Entity-based cross-document coreferencing using the vector space model. In: Proceedings of the 17th international conference on computational linguistics, vol. 1. Stroudsburg: Association for Computational Linguistics; 1998. P. 79–85
4. Barabási AL, Albert R. Emergence of scaling in random networks. Science 1999; 286(5439):509–12
5. Blondel VD, Guillaume JL, Lambiotte R, Lefebvre E. Fast unfolding of communities in large networks. J Stat Mech: Theory Exp 2008; 2008(10): P10,008
6. Brandes U, Delling D, Gaertler M, Gorke R, Hoefer M, Nikoloski Z, et al. On modularity clustering. IEEE Trans Knowl Data Eng 2008;20(2):172–88
7. Chakraborty T, Sikdar S, Ganguly N, Mukherjee A. Citation interactions among computer science fields: a quantitative route to the rise and fall of scientific research. Soc Netw Anal Min 2014;4(1):1–18
8. Chakraborty T, Sikdar S, Tammana V, Ganguly N, Mukherjee A. Computer science fields as ground-truth communities: their impact, rise and fall. In: Proceedings of advances in social networks analysis and mining (ASONAM). New York: ACM, 2013. P. 426–33
9. Chakraborty T, Srinivasan S, Ganguly N, Mukherjee A, Bhowmick S. On the permanence of vertices in network communities. In: Proceedings of the 20th ACM SIGKDD international conference on knowledge discovery and data mining, KDD 2014. New York, NY: ACM; 2014. P. 1396–405
10. Clauset A, Newman M, Moore C. Finding community structure in very large networks. Phys Rev E 2004;70(6). http://journals.aps.org/pre/abstract/10.1103/PhysRevE.70.066111
11. Corneil DG, Dalton B, Habib M. LDFS-based certifying algorithm for the minimum path cover problem on cocomparability graphs. SIAM J Comput 2013;42(3):792–807
12. Corneil DG, Krueger RM. A unified view of graph searching. SIAM J Discr Math 2008;22(4):1259–276
13. Creusefond J, Largillier T, Peyronnet S. Finding compact communities in large graphs. In: Proceedings of advances in social networks analysis and mining (ASONAM), 2015. ACM; 2015. P. 1457–464
14. Creusefond J, Largillier T, Peyronnet S. On the evaluation potential of quality functions in community detection for different contexts. In: Advances in network science. Springer; 2016. P. 111–125
15. Flake GW, Lawrence S, Giles CL. Efficient identification of Web communities. In: Proceedings of the sixth ACM SIGKDD international conference on Knowledge discovery and data mining. New York: ACM, 2000. P. 150–60
16. Fortunato S. Community detection in graphs. Phys Rep 2010;486(3–5):75–174
17. Fortunato S, Barthelemy M. Resolution limit in community detection. Proc Natl Acad Sci 2007;104(1):36–41
18. Girvan M, Newman ME. Community structure in social and biological networks. Proc Natl Acad Sci 2002;99(12):7821–826
19. Hansen P, Jaumard B. Minimum sum of diameters clustering. J Class 1987;4(2):215–26
20. Hu Y. Efficient, high-quality force-directed graph drawing. Math J 2005;10(1):37–71
21. Kannan R, Vempala S, Vetta A. On clusterings: good, bad and spectral. J ACM (JACM) 2004;51(3):497–515
22. Klimt B, Yang Y. Introducing the enron corpus. In: CEAS. 2004
23. Lancichinetti A, Fortunato S, Kertész J. Detecting the overlapping and hierarchical community structure in complex networks. New J Phys 2009;11(3):033015

24. Leskovec J, Kleinberg J, Faloutsos C. Graph evolution: densification and shrinking diameters. ACM Trans Knowl Discov Data 2007;1(1):2
25. Leskovec J, Lang KJ, Dasgupta A, Mahoney MW. Statistical properties of community structure in large social and information networks. In: Proceedings of the 17th international conference on World Wide Web. ACM; 2008. P. 695–704
26. Leskovec J, Lang KJ, Mahoney M. Empirical comparison of algorithms for network community detection. In: Proceedings of the 19th international conference on World wide web. ACM; 2010. P. 631–40
27. Leskovec J, Mcauley JJ. Learning to discover social circles in ego networks. In: Advances in neural information processing systems; 2012. P. 539–47
28. Mislove A, Marcon M, Gummadi KP, Druschel P, Bhattacharjee B. Measurement and analysis of online social networks. In: Proceedings of the 5th ACM/Usenix internet measurement conference (IMC 2007), San Diego, CA; 2007
29. Newman ME, Girvan M. Finding and evaluating community structure in networks. Phys Rev E 2004;69(2):026113
30. Pons P, Latapy M. Computing communities in large networks using random walks. J Graph Algorithms Appl 2006;10(2):191–218
31. Radicchi F, Castellano C, Cecconi F, Loreto V, Parisi D. Defining and identifying communities in networks. Proc Natl Acad Sci USA 2004;101(9):2658–2663
32. Raghavan U, Albert R, Kumara S. Near linear time algorithm to detect community structures in large-scale networks. Phys Rev E 2007;76(3). http://journals.aps.org/pre/abstract/10.1103/PhysRevE.76.036106
33. Rosvall M, Bergstrom CT. Maps of random walks on complex networks reveal community structure. Proc Natl Acad Sci 2008;105(4):1118–123
34. Seidman SB. Network structure and minimum degree. Soc Netw 1983;5(3):269–87
35. Šubelj L, Bajec M. Model of complex networks based on citation dynamics. In: Proceedings of the 22nd international conference on World Wide Web; 2013. P. 527–30
36. Tarjan RE. Efficiency of a good but not linear set union algorithm. J ACM (JACM) 1975;22(2):215–25
37. Traag VA, Krings G, Van Dooren P. Significant scales in community structure. Sci Rep 2013;3. http://www.nature.com/articles/srep01060?WT.ec_id=SREP-631-20130201 and http://www.nature.com/articles/srep02930
38. van Dongen S. Graph clustering by flow simulation. Ph.D. thesis (2000)
39. Van Laarhoven T, Marchiori E.: Axioms for graph clustering quality functions. J Mach Learn Res 2014;15(1):193–215
40. Watts DJ, Strogatz SH. Collective dynamics of 'small-world' networks. Nature 1998; 393(6684):440–42
41. Yang J, Leskovec J. Defining and evaluating network communities based on ground-truth. Knowl Inf Syst 2012;42(1):81–213

Computational Data Sciences and the Regulation of Banking and Financial Services

Sharyn O'Halloran, Marion Dumas, Sameer Maskey, Geraldine McAllister, and David K. Park

1 Introduction

This paper combines observational methods with computational data science techniques to understand the design of financial regulatory structure in the USA. The centerpiece of the analysis is a database encoding the text of financial regulation statutes from 1950 to 2010. Among other variables, we identify the amount of discretionary authority Congress delegates to executive agencies to regulate the banking and financial services sector. The analysis requires aggregating measures from thousands of pages of text-based data sources with tens of thousands of provisions, containing millions of words. Such a large-scale manual data tagging project is time consuming, expensive, and subject to potential measurement error.

To mitigate these limitations, we employ Natural Language Processing (NLP), such as parsing, Machine Learning (ML) techniques, such as Naive Bayes and feature selection models, and topic modeling, to complement the observational study. While none of these techniques alone are unique, the combination of these techniques applied to financial regulation is unique. These methods allow us to efficiently process large amounts of complex text and represent them as feature vectors, taking into account a law's topic, words, and phrases. These feature vectors can be easily paired with predefined policy attributes specified in the manual coding.

S. O'Halloran (✉) • S. Maskey • G. McAllister • D.K. Park
Columbia University New York, NY, USA
e-mail: so33@columbia.edu; srm2005@columbia.edu; gam2116@columbia.edu; dkp7@columbia.edu

M. Dumas
Santa Fe Institute, Santa Fe, NM, USA
e-mail: marion@santafe.edu

© Springer International Publishing AG 2017
M. Kaya et al. (eds.), *From Social Data Mining and Analysis to Prediction and Community Detection*, Lecture Notes in Social Networks,
DOI 10.1007/978-3-319-51367-6_8

The purpose of this paper is to analyze how these methods can be used to build predictive models of financial regulation from the text of the laws.

In a previous paper, O'Halloran et al. [35] analyze how traditional observational studies could be enhanced by computational analysis. The results show that feature selection models that combine observational methods with computational data science techniques greatly improve the accuracy of the measurements. We expand the previous analysis by computationally assigning laws to categories via topic modeling to further enhance the accuracy of the predictive model over feature selection models alone. Furthermore, we show how these techniques can be used to develop robust standard errors and thereby facilitate hypothesis testing about the design of financial regulations.

The analysis provides policymakers with a tool to combine the insights of subject matter experts with the advantages of computational analysis of text to score financial regulation laws to understand their impact on market performance. This paper thereby offers a new path, illustrating how triangulating different methods can facilitate the measurement of otherwise expensive and difficult to code institutional variables. This in turn furthers our understanding of important substantive public policy concerns.

The paper proceeds in the following steps. Part 2 sets the stage by presenting the illustrative example that structures the subsequent analysis: how to test hypotheses derived from the political economy of regulatory design and more specifically financial market regulation. We review the main hypotheses in the field and demonstrate the traditional observational approach to measurement by detailing the coding method used to construct the financial regulation database. We then discuss the limits of such observational approaches and how advances in computational data sciences can mitigate some of these shortcomings. Part 3 presents the computational methods used in the paper: a combination of NLP to parse all the documents of the financial regulation laws into feature vectors, topic modeling to cluster laws into relevant policy sub-domains, and finally supervised models to predict the outcome variable of interest. Part 4 presents our results, applying the ML algorithms to compare the power of alternative models in predicting regulatory discretion. Part 5 discusses the significance of the findings in light of the research design challenges discussed earlier. Conclusions and future developments close the paper.

2   Measurement and Inference in Testing Theories of Financial Market Regulation

2.1   The Why and How of Financial Regulation

What explains the structure of financial regulation? Where, how, and by whom policy is made significantly impacts market outcomes.[1] When designing financial

---

[1] A number of studies show that government institutions matter for the regulation of markets. Keefer [20] argues that competitive governmental structures are linked with competitive markets.

regulation laws, Congress specifies the rules and procedures that govern bureaucratic actions. The key is how much discretionary decision making authority Congress delegates to regulatory agencies. In some cases, Congress delegates broad authority, such as mandating the Federal Reserve to ensure the "safety and soundness of the financial system." Other times, Congress delegates limited authority, such as specifying interest rate caps on bank deposits.

A recurring theme in the political economy literature of regulatory design is that the structure of policy making is endogenous to the political environment in which it operates.[2] Epstein and O'Halloran [10] show that Congress delegates policymaking authority to regulatory agencies when the policy preferences of Congress and the executive are closely aligned, policy uncertainty is low, and the cost (political and otherwise) of Congress setting policy itself is high. Conflict arises because of a downstream moral hazard problem between the agency and the regulated firm, which creates uncertainty over policy outcomes.[3]

Application of these theoretical insights to financial regulation is well motivated. Banking is a complex policy area where bureaucratic expertise is valuable and market innovation makes outcomes uncertain. Morgan [31], for instance, shows that rating agencies disagree significantly more over banks and insurance companies than over other types of firms. Furthermore, continual innovation in the financial sector means that older regulations become less effective, or "decay," over time. If it did not delegate authority in this area, Congress would have to continually pass new legislation to deal with new forms of financial firms and products, which it has shown neither the ability nor inclination to do.[4] Overall, then, we have the following testable hypotheses: Congress delegates more discretion when: (1) The preferences of the President and Congress are more similar; and (2) Uncertainty over market outcomes (moral hazard) is higher.

In particular, separation of powers and competitive elections are correlated with strong investor protection and lending to the private sector. Barth et al. [4] show countries that encourage private enforcement of banking laws and regulation (e.g., through litigation) rather than direct control or no regulation at all, have the highest rates of financial sector development and therefore capital formation. Historical studies of financial development in the USA tell similar stories. Kroszner and Strahan [21] show that the relative political strength of winners from deregulation (large banks and smaller, bank-dependent firms) and the losers (small banks and insurance firms) explains the timing of bank branching deregulation across states in the USA. Haber [18] argues that governments free from outside political competition will do little to implement regulations in the banking sector.

[2]For early work in this area, see, for example, McCubbins and Schwartz [24] and McCubbins et al. [25, 26].

[3]Excellent technical work on the optimal type of discretion to offer agencies is provided by Melumad and Shibano [27] and Alonso and Matouschek [3], and Gailmard [11]. A series of studies examine the politics of delegation with an executive veto [41], civil service protections for bureaucrats [12, 13], and executive review of proposed regulations [43], among others. See also Bendor and Meirowitz [5] for contributions to the spatial model of delegation and Volden and Wiseman [42] for an overview of the development of this literature.

[4]Maskin and Tirole [22] and Alesina and Tabellini [2] also emphasize the benefits of delegation to bureaucrats and other non-accountable officials.

Groll et al. [16] expand on this work, addressing whether policymakers regulate financial markets on their own or delegate regulatory authority to government agencies when faced with uncertainty about firm-specific investments and systemic risk at the financial services level. The executive is better informed and knows the exact correlation but puts greater weight on the social cost of a possible bailout.[5]

They conclude that Congress delegates regulatory authority when (1) the preferences of the executive and Congress are more similar, (2) the costs of a bailout are high, (3) there is more uncertainty about investment risk and systemic risk, and (4) Congress's bailout concern is low relative to the executive's. Further, financial services are more heavily regulated when firm-specific investments and systemic risks are uncertain. But when interbranch preferences differ or perceived systemic risk is low, Congress may allow risky investments to be made that, *ex post*, it wished it had regulated.

To illustrate the trade-off between policy differences and market uncertainty, the analysis focuses on Congress's and the executive's preferences and information differences. Congress prefers, in general, lower regulation thresholds than the executive because it puts less weight on the cost of a possible bailout. This is referred to as the preference difference between Congress and the executive, which is part of Congress's trade-off between the information advantage of the executive and the difference in preferred policy.

Figure 1 illustrates the implications when Congress would regulate on its own and when it would delegate to the executive. The shaded area indicates situations in which Congress delegates, while outside this area Congress makes policy on its own. Firms do not internalize a potential systemic failure in the absence of

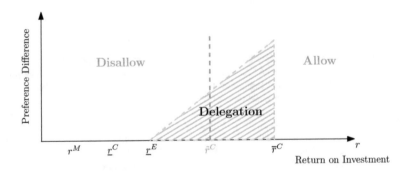

Fig. 1 Regulation and delegation

---

[5]The argument is quite intuitive: When the financial system experiences a shock, then constituents are more likely to hold the president and the executive accountable than any individual member of Congress. For formal proofs of these propositions, the reader is referred to Groll et al. [16]. A similar argument is made in trade policy as constituents hold the president and the executive more accountable for the overall economic conditions, which explains more free-trade oriented positions by the executive than Congress. See, for example, O'Halloran [34].

regulation and would make any investment that yields them a nonnegative expected return—that is, any investments at or above r^M in Fig. 1. When Congress regulates, then any investments that would yield Congress a negative expected social return including the cost of a possible bailout—that is, any returns below r^C in Fig. 1— would be banned and all others would be allowed. When Congress delegates, then, the executive sets a requirement in a similar spirit: banning investments with a negative expected social return that accounts for the cost of a possible bailout and the executive's salience. The executive knows the correlation and can therefore set a standard for uncorrelated investments, $\underline{r}^E$ and for correlated investments that of Congress's preferred threshold of $\bar{r}^C$ because of limited discretion from Congress. The delegation to the executive with discretion follows then from these two thresholds. All three regulatory standards are increasing in (1) the actual cost of a bailout, (2) the salience of a possible bailout, and (3) the perceived likelihood of correlated investments. But the stringency decreases when the likelihood of successful investments increases.

Focusing on the preference difference and expertise, the executive puts relatively more weight on the bailout cost than Congress, and therefore Congress values the executive's expertise most when there is no preference difference and delegates more discretion to the executive. But as the executive puts an increasing weight on the bailout cost, or Congress a decreasing weight, the preference difference increases and Congress delegates less to the executive because of its higher standards, which would imply a movement along the vertical axis in Fig. 1. In other words, the delegation area shrinks as the disagreement between Congress and the executive increases and Congress prefers to regulate on its own. However, when the costs of a potential bailout increase or Congress's uncertainty over correlated investments increases, the area of delegation expands as Congress gains relatively more from the executive's expertise increase.

The preference difference also has implications for investments that are highly risky but high-returning when Congress perceives a low likelihood of correlated investments. In such situations, Congress prefers to regulate the financial investments on its own—that is, Congress would set a return requirement that is actually below the executive's standard for uncorrelated investments $\underline{r}^E$ in Fig. 1. The reason is that the executive's expertise about correlated investments is not expected to be valuable to Congress, and the executive's standard is perceived as too stringent given the preference difference.

Overall, then, we have the following testable hypotheses:

1. Congress delegates more discretion when:

   (a) The policy preferences between Congress and the executive become more similar;
   (b) Firms' investment risks become more uncertain, and;
   (c) There is more uncertainty about investment risk; and
   (d) The costs of a bailout are higher.

2. The more Congress cares about bailout costs, the higher are

(a) Congress's preferred level of regulation,
(b) The executive's discretion and regulation, and
(c) Overall levels of regulation.

2.2 Financial Regulation Laws as Data

The previous section established that we need to test the hypothesis that regulatory design responds to the political preferences of Congress and the executive. Testing such hypotheses is challenging because it requires measuring discretionary authority. Measuring policy or institutional features is generally difficult, because these variables have no intrinsic scale and are not directly observable, arising instead from a combination of many rules. These rules are qualitative and thus require parsing texts. In our motivational example, the key variable of interest is the amount of discretionary authority Congress delegates to regulatory agencies to set policy (Discretion Index). It depends not only on the amount of authority delegated (Delegation Ratio) but also on the administrative procedures that constrain executive actions (Constraint Ratio). In what follows, we explain the process used to construct a measure of agency discretion. This process illustrates how measurement often proceeds in the social sciences, absent the help of computational tools.

We create a new database comprising all U.S. federal laws enacted from 1950 to 2010 that regulate the financial sector. The unit of analysis is an individual law, which specifies the rules and producers that regulate the actions of financial market participants. The database contains 120 public laws. The average corpus of text of a legislative summary is 6,278 words.[6] Because the Discretion Index is a combination of the Delegation Ratio and the Constraint Ratio, we present the measurement as a two-step process.

Delegation Ratio Delegation is defined as authority granted to an executive branch actor to move policy away from the status quo.[7] For each law, we code if substantive authority is granted to executive agencies, the agency receiving authority (for example, the U.S. Securities and Exchange Commission, Treasury, etc.), and the location of the agency within the administrative hierarchy (for example, independent agency, cabinet, etc.).

---

[6]The analysis relies on legislative summaries provided by *Congressional Quarterly* and contained in the Library of Congress's Thomas legislative database.

[7]For example, the Dodd–Frank Wall Street Reform and Consumer Protection Act of 2010 (Dodd–Frank Act) delegated authority to the Federal Deposit Insurance Corporation to provide for an orderly liquidation process for large, failing financial institutions.

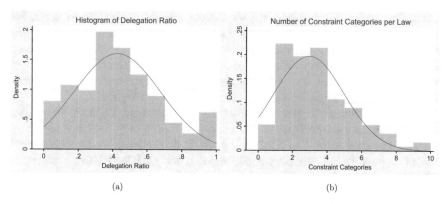

Fig. 2 Delegation and constraint ratios (**a**) Delegation ratio (**b**) Constraint ratio

To measure delegation, each law in our database was manually read independently, its provisions numbered, and all provisions that delegated substantive authority to the executive branch were identified and counted.[8]

From these tallies, we calculate the delegation ratio by dividing the number of provisions that delegate to the executive by the total number of provisions. In the database, each law contains an average of 27 provisions of which 11 delegate substantive authority to four executive agencies. The average delegation ratio across all laws then is 0.41 or 11/27. The histogram of delegation ratios is shown in Fig. 2a. As indicated, the distribution follows a more or less normal pattern, with a slight spike for those laws with 100% delegation (these usually had a relatively small number of provisions).

Constraint Ratio Executive discretion depends not only on the amount of authority delegated but also on the administrative procedures that constrain executive actions. Accordingly, we identify 14 distinct procedural constraints associated with the delegation of authority and note every time one appears in a law.[9] Including all 14 categories in our analysis would be unwieldy, so we investigated the feasibility of using principal components analysis to analyze the correlation matrix of constraint categories. As only one factor was significant, first dimension factor scores for each law were calculated, converted to the [0,1] interval, and termed the constraint index. Each law on average contained three constraints of the possible 14, yielding an overall constraint ratio of 0.21. The histogram of constraint ratios is shown in Fig. 2b

---

[8]To ensure the reliability of our measures, each law was coded independently by two separate annotators. It was reviewed by a third independent annotator, who noted inconsistencies. Upon final entry, each law was then checked a fourth time by the authors. O'Halloran et al. [35] provide a detailed description of the coding method used in the analysis.

[9]Examples of procedural constraints include spending limits, and legislative action required, etc. See O'Halloran et al. [35] for a detail description of these constraints.

Discretion Index From these data, we calculate an overall discretion index. For a given law, if the delegation ratio is D and the constraint index is C, both lying between 0 and 1, then total discretion is defined as $D * (1 - C)$—that is, the amount of unconstrained authority delegated to executive actors.[10] The more discretion an agency has to set policy, the greater the leeway it has to regulate market participants. Lower levels of agency discretion are associated with less regulation.

As an illustration for how this measure is calculated, the Dodd–Frank Act contains 636 provisions of which 314 delegate authority to 46 executive agencies, yielding a delegation ratio of 0.5. The law also indicated ten procedural constraints out of a possible 14, yielding a constraint index of 0.7 (10/14). Combining delegation and constraints ratios produces a discretion index of $0.5 * (1 - 0.7) = 0.1$.

To verify the robustness of our estimates and confirm that our choice of aggregation methods for constraints does not unduly impact our discretion measure, Fig. 3 shows the average discretion index each year calculated four different ways. As the time series patterns are almost identical, our choice of method number four (continuous factors, first dimension) is not crucial to the analysis that follows.

As a basic check on our coding of delegation and regulation, we compare the distribution of the discretion index for laws that regulated the financial industry overall, and laws that deregulated. We would expect that laws regulating the industry would delegate more discretionary authority, and Fig. 4 shows that this is indeed the

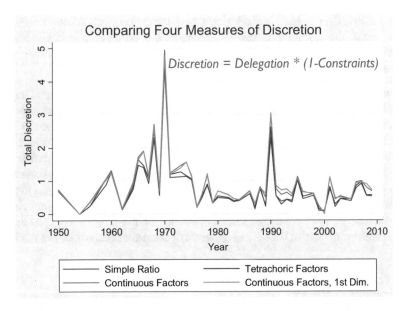

Fig. 3 Four measures of executive discretion

---

[10]See Epstein and O'Halloran [10] for a complete discussion of this measure.

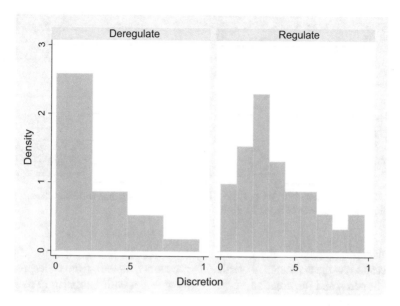

Fig. 4 Distribution of discretion index for laws that deregulate and laws that regulate

case. The average discretion index for the 24 laws that deregulate is 0.24, as opposed to 0.35 for the 85 laws that regulate, in line with the hypotheses discusses above.[11]

Do differences in executive discretion, in turn, affect the financial industry? It is in general difficult to determine a measure of the degree to which regulation is successful, but Philippon and Reshef's [36] measure of excess wages in the financial industry serves as a useful proxy; the greater the degree of regulation, the lower excess wages should be.

Accordingly, Fig. 5a overlays total discretion delegated each year with Phillipon and Reshef's excess wage measure. The trends seem to be in the correct direction: excess wages began to increase in the 1960s, then a spate of regulation drove them down. They increased again the late 1980s, and again were lowered by a spike in regulation. Finally wages started rising precipitously during the 1990s and on to the first decade of the 2000s, but there was no corresponding rise in regulation to meet them, so they just kept increasing to the levels we see today. The trends in Fig. 5a thus accord with the notion that delegating discretionary authority to executive branch officials does constrain the level of excess wages in the financial industry.[12]

---

[11]The remaining laws neither regulated nor deregulated.

[12]Further, the patterns also seem consistent with the notion that regulations "decay" over time as new financial instruments appear to replace the old one. If one estimates a Koyck distributed lag model $y_t = \alpha + \beta x_t + \beta \phi x_{t-1} + \beta \phi^2 x_{t-2} + \cdots + \epsilon_t$ via the usual instrumental variables technique, then $\beta = -0.025$ and $\phi = 0.49$, indicating that regulations lose roughly half their effectiveness each year. See Wooldridge [45], pp. 635–637 for details of the estimation technique.

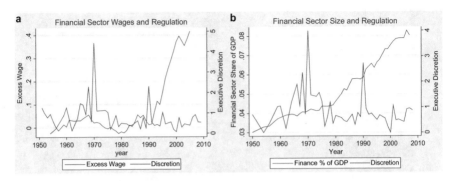

Fig. 5 Regulatory discretion and the financial sector (**a**) Discretion and financial wages (**b**) Discretion and importance of financial sector

They also pose a puzzle: why was the spate of financial innovation in the 1960s—this decade saw the explosion of credit in the economy, including the widespread use of credit cards and the creation of credit unions—met with a regulatory response, while the most recent innovations—derivatives, non-bank lenders and the rise of the shadow banking system—were not? We postpone our suggested answer until the concluding section of this paper.

Figure 5a also indicates that the trend in recent decades has been for Congress to give executive branch actors less discretion in financial regulation. Since the Great Society era of the 1960s, and on into the early 1970s, the total amount of new executive branch authority to regulate the financial sector has generally declined. The exceptions are a few upticks in discretion which coincide with the aftermaths of well-publicized financial crises and scandals, including the Savings and Loan crisis, the Asian crisis, and the Enron scandal. Otherwise, the government has been given steadily less authority over time to regulate financial firms, even as innovations in that sector have made the need for regulation greater than ever, and even as the importance of the financial sector in the national economy has greatly increased as illustrated in Fig. 5b.

What is the source of this decrease in discretion? As shown in Fig. 6a, the amount of authority delegated to oversee the financial sector has remained fairly constant over time, perhaps decreasing slightly in the past decade. The trends in Fig. 5, then, are due mainly to a large and significant increase in the number of constraints placed on the regulators' use of this authority. In addition, we find that the number of actors receiving authority has risen significantly over the time period studied, as also shown in Fig. 6a. However, the location of these agencies in the executive hierarchy has changed as well, away from more independent agencies to those more directly under the president's control as illustrated in Fig. 6b.

Finally, we investigated the impact of this changed regulatory structure on overall market performance, statistically analyzing the impact of greater agency discretion on yearly changes in the Dow Jones Industrial Average and on the number and

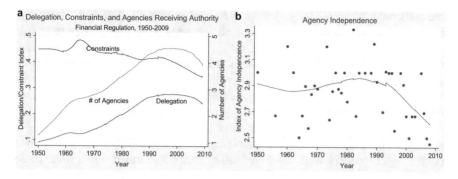

Fig. 6 Trends in constraints and delegation (**a**) Delegation, constraints, and number of agencies (**b**) Independence of agencies

size of bank failures. We found, somewhat surprisingly, that financial markets react positively to higher levels of regulation. The Dow Jones average was half a percent higher, on average, in years in which Congress gave executive actors authority to regulate financial institutions, as compared to years in which no authority was given. Furthermore, the number and size of bank failures decreased significantly following regulatory reform.[13]

Overall, then, our preliminary analysis suggests that the current rules defining financial regulation may create a web of interlocking and conflicting mandates, making it difficult for regulators to innovate the rules and standards governing the financial industry, while at the same time opening regulatory agencies to industry capture. The problem is not lack of regulation, then, but that regulators have little discretion. Modern laws delegate less, constrain more, and split authority across more agencies than their predecessors. This has created a situation where many areas of financial activity are heavily regulated by the Federal government, but those charged with oversight are hamstrung by overlapping jurisdictions, the need for other actors to sign off on their policies, or outright prohibitions on regulatory actions by Congress.

2.3   Limitations of the Observational Method

Section 2.2 defined the discretion index as $D * (1 - C)$, where D is the delegation ratio and C is the constraint index. As noted earlier, the process discussed in Sect. 2.2 is a standard political economy approach to measurement for the testing of hypotheses. The above analysis adopts a research design based on observational

---

[13]See Groll et al. [16].

methods, which potentially suffer from a number of well-known shortcomings. First, observational studies assume that all variables of interest can be measured according to a pre-specified coding rule. For example, the analysis posits that discretion can be calculated as a combination of delegation and constraints. In constructing these measures, the coding rules invariably impose a structure on the text, indicating some words or phrases as delegation and others as constraints. This can lead to coding bias.

Second, this approach can lead to substantial coding error, especially as it is scaled to larger datasets. Indeed, annotating the original data is extremely time consuming, especially when derived from disparate text-based sources, as we do here. The resources needed to extract the appropriate information, train annotators, and code the data can prove prohibitive and is prone to error. For example, consider again the Dodd–Frank Act, which covers the activities financial institutions can undertake, how these institutions will be regulated, and the regulatory architecture itself. Recall the law contains 636 major provisions, of which 314 delegate authority to some 46 federal agencies. In addition, the Act has a total of 341 constraints across 11 different categories, with 22 new agencies created. If we process the text of this law by the coding method detailed above, data annotators, trained in political economy theories, would read and code the provisions based on the rulebook provided. In effect, coders would have to read 30,000 words—the length of many novels. Unlike novels, however, legislation is written in complex legal language, which must be interpreted correctly and in painstaking detail. Consequently, there is the possibility that data annotators will introduce noise when coding laws. Measurement error is in some way inevitable, but the rule-based approach to coding prevents us from quantifying the uncertainty arising from these errors.

Third, standard econometric techniques, upon which many political economic studies rely, including the one conducted here, face difficulty in analyzing high-dimensional variables that could theoretically be combined in a myriad of ways. For example, Fig. 3 shows four possible alternatives to calculate the discretion index by varying the weights assigned to the different categories of procedural constraints.

In the following section, we explore data science methods and identify the techniques best suited to address the limitations of traditional observational methods. The computational methods presented below provide potential improvements over manual coding from a set of defined rules, including:

- We are not limited by the amount of data we can process.
- We are not limited to a handful of coding rules to quantify each law for building the discretion model.
- We can take into account the raw text of the law to explore word combinations, syntactic and dependency relations, and identify other sets of features that otherwise would be difficult to encode manually.

Table 1 provides a comparison of observational methods and data science techniques along three criteria: coding legislation, analysis, and internal validity. The overview illustrates the shortcomings of using only manual, rules-based coding methods and the way these new methods can enhance observational studies. In sum,

Table 1 Comparison of observational study and new machine learning method

	Observation study		Machine learning	
	Process	Disadvantages	Process	Advantages
Coding	Coding rules	Labor and time costs, and coding error in large corpus	Natural language processing	Efficiency and consistency
	Human coders	Coding bias	Data can be words, semantic units, relations, data structure	Detection of implicit/laten factors
	Checking consistency	Measurement uncertainty not quantified	Validation on test set or cross-validation	Quantification of measurement uncertainty
Analysis	Hypothesis testing	Number of hypothesis	Various Naive Bayes Models	Not limited by data amount
	Regression analysis	Number of variables	Comparing model accuracies	Optimization of complexity
	Correlation on variables	Scaling and sensitive to outliers	Comparison with human coding	Flexibility beyond coding rules
Internal validity	Low	Missing underlying structure	High	No functional form
	Panoply of variables in single analysis	Missing important variables	Analysis of words, relations, semantic dependencies	Low generalization errors
		Imposed functional form		No overfitting

computational analysis helps lessen error, quantify the uncertainty arising from these errors, find additional variables and patterns (data features), and improve the predictive power of models.

It is important to keep in mind that, while NLP and ML techniques can analyze extensive text-based data to test theories of policymaking and regulatory design, they rely on the critical data and hypotheses initially produced by subject matter experts to inform or "seed" the model and train complex algorithms. Therefore, data science techniques can be seen as a complement to observational studies and theoretical analysis.

3   Methods: Computational Coding of Financial Regulation Laws

We next describe the computational models used to predict the level of agency discretion. We show that both unsupervised and supervised algorithms can be combined to provide sparse representations of large datasets of laws and build predictive models of a statute's regulatory structure. In particular, we seek to determine what factors or "features" of a law predict agency discretion and also build a model that predicts discretion with high accuracy. Identifying the key features, words or word patterns, that predict the level of agency discretion in a given law helps refine and develop better proxies for institutional structure. We will compare how different types of features (computationally selected features, observational features and topics identified as latent variables in an unsupervised model) each contribute to increasing the accuracy of the predictive model. Our analysis thereby offers a novel approach to analyzing institutional design. Only recently have data science techniques been applied to study financial regulation and public policy more broadly.

First, however, we need to represent the passages of the legal documents in a format that is suitable for ML methods. We employ NLP techniques to convert the text of the laws into feature vectors. Some of the many different ways to encode text into features are listed here:

- Bag of Words: A bag of words model represents text as a feature vector, where each feature is a word count or weighted word count (McCallum and Nigam [23].
- Tag Sequences: Sentences or chunks of text are tagged with various information, such as Parts of Speech (POS) tags [8] or Named Entities (NEs, see Nadeau and Sekine [32]), which can be used to further process the text.
- Graphs: Documents or paragraphs of the documents can be represented in graphs, where nodes can model sentences, entities, paragraphs, and connections represent relations between them [28].
- Semantic Representation: A sequence of words mapped into an organized structure that encodes semantics of the word sequences [15].

These methods can be applied to represent text, thereby allowing machines to extract additional information from the words (surface forms) of the documents. Depending on the problem being addressed, one or more of these tools may be useful. We next explain the representation form adopted for the computational experiments below.

3.1 Data Representation Using Natural Language Processing

We must represent each individual law in a form suitable for ML algorithms to take as inputs. We first convert the raw text of an individual law in feature representation format. For the current experiment, we convert the text of the financial regulation laws to Word Vectors. We describe the process of converting text into feature vectors below.

Step 1: Data Cleaning—For each law, we first clean the text to remove any words that do not represent core content, including meta information such as dates, public law (P.L.) number and other metadata that may have been added.

Step 2: Tokenization— After cleaning the data, we tokenize the text. Tokenization in NLP involves splitting a block of text into a set of tokens (words). This involves expanding abbreviations *(Mr. > Mister)*, expanding words *(I've > I have)*, splitting punctuation from adjoining words *(He said, > He said ,)* and splitting text using a delimiter such as white space *(bill was submitted > [(bill) (was) (submitted)])*.

Step 3: Normalization—Once tokenized, we must then normalize the data. The normalization of data involves having consistent tokenization across the same set of words.

Step 4: Vocabulary—In order to represent text in the form of feature vectors we must find the total vocabulary of the corpus appended with the additional vocabulary of the language.

Step 5: Vector Representation—Once we have defined the vocabulary, we can treat each word as adding one dimension in the feature vector that represents a block of text.

Let d_i be the document i. Let $y = w_1, w_2, \ldots, w_n$ be the vector representation of that document d_i, where w_k represent the existence of word w_k in the document d_i. Let us take an example piece of text from the Dodd–Frank Act, contained in section 1506.

$d_i=$"..the definition of core deposits for the purpose of calculating the insurance premiums of banks". Let n be the total vocabulary size. The vector representation y for this document d_i will consist of a vector of length n where all values are set to zero except for the words that exist in document d_i. The total vocabulary size n tends to be significantly bigger than the number of unique words that exist in a given document so the vector tends to be very sparse. Hence, the vector y for document d_i is stored in sparse form such that only non-zero dimensions of the vector are actually stored. The vector of d_i will be

$$y = \{\text{definition} = 1.0, \text{representation} = 1.0, \text{core} = 1.0, \text{purpose} = 1.0, \text{calculate} = 1.0,$$

$$\text{insurance} = 1.0, \text{premium} = 1.0, \text{bank} = 1.0\}.$$

This is a binary vector representation of the text d_i. Given that the word is present in the document, we can in fact keep track of the word count in the given document d_i and store counts in the vector rather than storing the binary number representing it. Correspondingly, this generates a multinomial vector representation of the same text. If we take the entire Dodd–Frank Act as d_q, rather than sample text, and store counts for each word, we yield the vector representation of the Act as:

$$y = \{sec = 517.0, financial = 304.0, securities = 106.0, requires = 160.0, federal = 154.0,$$
$$requirements = 114.0, \ldots, inspection = 2.0\}.$$

Step 6: *TF $*$ IDF* **Transformation**—Once we represent the document in raw word vector format, we can improve the vector representation format by weighting each dimension of the vector with a corresponding term known as *Inverse Document Frequency* (IDF) [39]. An IDF transformation takes into account giving less weight to words that occur across all documents. For example, if the word *SEC* occurs frequently in all laws, then the word *SEC* has less distinguishing power for a given class than *house*, which may occur less frequently, but is strongly tied to a given class. We re-weight all the dimensions of our vector d_q by multiplying them with the corresponding IDF score for the given word. We can obtain IDF scores for each word w_j by creating an IDF vector that can be computed by Eq. (1).

$$\text{IDF}(w_j) = \log\left(\frac{N}{\#count - of - Doc - with - w_j}\right), \tag{1}$$

where N is the total number of documents in the corpus and $\#count - of - Doc - with - w_j$ is the total number of documents with the word w_j. If the word w_j occurs in all documents, then the IDF score is 0.

3.2 Unsupervised Model: Topical Modeling of Financial Regulation Laws

Unsupervised models can help discover underlying clusters in the data, such as groups of documents that address the same issue. In the context of our analysis, clustering is particularly useful because we expect that the discretion level will vary systematically by policy issue, depending on the level of political risk (see Sect. 2.1). This prompts us to explore ways of clustering our corpus of financial regulation laws according to finer policy domains in order to include these policy domains as features in the supervised model described in the next section. More generally, the clustering of laws by policy domains should prove useful in new and larger corpora of laws as a way to describe substantive issues covered by the laws. Clustering can also facilitate the coding of institutional features, as the relevance of different institutional features can vary depending on the policy domain (e.g., health versus financial regulation).

Different models exist to describe corpora as a set of clusters or topics, in which topics are distributions over word frequencies or other scores such as TF-IDF scores. An example includes k-means [40]. Topic models is a family of models particularly well suited for this task. These models start from the same word vector representation of the document described earlier. They then model this word vector as a distribution over topics (for example, a financial law can be about both securities and commodities) and topics as distributions over words. Conversely, one can think of each word of a document as belonging to a topic and each document being a mixture of multiple topics, a realistic description for many documents, including laws. One of the first and most widely used topic models is the Latent Dirichlet Allocation model (LDA), proposed by Blei et al. [7], of which other topic models are close variants. We thus use it to present the basic structure of topic models.

In a topic model with K topics, each topic k has a distribution β_k over words of the vocabulary. β_k is thus a vector of probabilities of observing each word, summing to 1. Each document d_i has a distribution over topics θ_i (a vector with elements $\theta_{i,k}$ for $k = 1, \ldots K$). $w_{i,j}$ is the observed word in the jth word position of document d_i, with topic assignment $z_{i,j}$ (the vector of topic assignment for the whole document being denoted z). The generative model for a document d_i with word vector y_i can be described by the following steps:

1. Draw the per topic proportions from a Dirichlet prior

$$\theta \sim \text{Dir}(\alpha)$$

2. For each of the n words in document d_i (where j is the jth word in the document d_i):

   (a) Draw the topic assignment

   $$z_{i,j} \sim \text{Multinomial}(\theta_i)$$

   (b) Draw the word given the topic assignment:

   $$w_{i,j} \sim \text{Multinomial}(\beta_{z_{i,j}})$$

The resulting log-likelihood for a document represented with word vector y_i is:

$$\log p(y_i | \alpha, \delta) = \log \int \left(\sum_{k=1}^{K} \prod_{j=1}^{n} p(w_j | z_{i,j} = k, \beta_k) p(z_{i,j} = k | \theta_i) \right) p(\theta_i | \alpha) d\theta_i$$

Different implementations exist to obtain the posterior distribution of the parameters of this model based either on sampling algorithms (most commonly Gibbs sampling as presented in Griffiths and Steyvers [14]) or variational expectation-maximization algorithms.

Many other models have been proposed on the basis of LDA. These models relax some of the assumptions in LDA and build into the basic model other features and dependencies to uncover more complex structures in documents (for an excellent recent review, see Blei [6]). Among others, dynamic topic models allow the distribution of topics and the content of topics to change over time or according to the documents' authors. A large suite of models integrate other types of meta-data. Bayesian non-parametric topic models uncover hierarchies of topics, from general to more fine-grained sub-topics. Correlated topic models allow correlations between topics (some topics are more likely to co-occur in a document than others).

In this project, we use the Latent Dirichlet Allocation model fitted by a variational expectation-maximization (VEM) algorithm as implemented in the `topicmodels` package in R (Grün and Hornik [17] an R interface to Blei's C implementation of VEM for fitting LDA). We also experimented with the Structural Topic Model (STM) of Roberts et al. [37], also fit with a VEM algorithm and implemented in the R package `stm`. STM is a topic-model developed with comparative political economy applications in mind. It incorporates document covariates as variables affecting the prevalence of topics in each document (for example, certain topics may be more prevalent in a Republican presidential speech than in a Democratic one) and the word content for each document (a Republican candidate may frame the same topic in different ways than a Democratic candidate). The generative model of the STM differs somewhat from LDA, since the per-document topic distribution is drawn from a logistic normal distribution (allowing covariates to influence the mean). The per-topic word distribution is drawn from an exponential distribution (also allowing the inclusion of covariates). The STM package is versatile, with many features to estimate models, select, explore, and visualize them. We will present the results from the STM in Sect. 4.5 and show that it successfully identifies relevant policy sub-domains, which additionally help improve the performance of our supervised models, to which we now turn.

3.3 Supervised Model: Naive Bayes

We frame our problem of predicting the level of agency discretion in a given law as a classification problem. The ML approach to supervised classification tasks is to train a model based on features of the data to predict observations' membership into the classes of interest labeled by a human. For this dataset, we denote the set of discretion classes as C_i, where i ranges from 0 to 5, the total number of classes used to tag individual laws for the *Level of Discretion*.

The Level of Discretion C_i in a given law is a subjective measure of how much discretionary authority is given to the agency in that law only. It is coded from

0 to 5, with 0 indicating that no discretionary authority was given to executive agencies to regulate financial markets and 5 meaning that significant discretionary authority was given. The *Discretion Level*, as a subjective measure, is different from the Discretion Index computed in Sect. 2.2. The latter index is derived from theory (based on the delegation ratio and constraint index, which are variables deemed salient on the basis of theoretical models of political economy). We use the *Discretion Level* instead of the *Discretion Index* because we want a measure that is independent from the coding rules dictated by theory, which could be wrong, and instead reflect a human's intuitive understanding of the text as a whole. Additionally, using subjective judgment as the *gold standard* that algorithms have to predict is a standard practice when ML models are built. With this in mind, let C_i be the level of discretion that we are trying to predict for a given document (law) y.

Many different machine learning algorithms are used in document/text classification problems. One of the most commonly applied algorithms is a Naive Bayes method. We build a Naive Bayes Model for predicting discretion level for each of the laws y.

We must compute $p(C_i|y)$ for each of the classes (discretion levels) and find the class C_i. $p(C_i|y)$ can be obtained by Eq. (2)

$$p(C_i|y) = \frac{p(C_i)p(y|C_i)}{p(y)} \qquad (2)$$

To find the best class C_i, we compute the argmax on the class variable:

$$i^* = \arg\max_i p(C_i|y). \qquad (3)$$

To compute $p(C_i|y)$, we use Bayes rule to obtain $p(C_i|y) = \frac{p(y|C_i)*p(C_i)}{p(y)}$. Since our task is to find argmax on C_i, we simply need to find C_i with the highest probability. As the term $p(y)$ is constant across all different classes, it is typically ignored. Next, we describe how we can compute $p(y|C_i)$ and $p(C_i)$.

$p(C_i)$ is the prior probability of class C_i. This term is computed on the training set by counting the number of occurrences of each class. In other words, if N is the total number of documents in training and N_i is the number of documents from class i, then $P(C_i) = \frac{N_i}{N}$.

In order to compute the probability $p(y|C_i)$, we assume that document y is comprised of the following words $y = \{w_1, w_2, \ldots, w_n\}$, where n is the vocabulary size. We make a conditional independence assumption that allows us to express $p(y|C_i) = p(w_1, \ldots, w_n|C_i)$ as

$$p(w_1, \ldots w_n|C_i) = \prod_{j=1}^{n} P(w_j|C_i). \qquad (4)$$

We compute $P(w_j|C_i)$ by counting the number of times word w_j appears in all of the documents in the training corpus from class C_i. Generally, *Add-one Smoothing* is used to address the words that never occur in the training document. Add-one smoothing is defined as follows: Let N_{ij} be the number of times word w_j is found in class C_i and let $P(w_j|C_i)$ be defined by Eq. (5), where n is the size of the vocabulary.

$$P(w_j|C_i) = \frac{N_{ij} + 1}{\sum_i N_{ij} + n} \tag{5}$$

Given a test document y, for each word w_j in y, we look up the probability $P(w_j|C_i)$ in this test document and substitute it into Eq. (5) to compute the probability of y being predicted as C_i. In Sect. 4, we describe the Naive Bayes Model built from different sets of features, thereby allowing us to compare the performance of our model in various settings.

4 Results: Comparing Models

As explained in Sect. 3.3, our purpose is to find characteristics of financial regulation laws that predict agency discretion. The computational analysis approach identifies policy features or attributes in the data that most accurately predict possible outcomes rather than tests hypothesizes about the impact of theoretically motivated explanatory variables on observed outcomes. In this section, we compare different versions of the Naive Bayes model, incrementally enriching and refining the feature set used to build the model. As noted in Sect. 3.3, to evaluate the performance of a given ML model in predicting agency discretion, each law is assigned a discretion level, ranging from 0 to 5, which serves as the target answer. We then compare the predictions yielded by each of the ML models against the baseline or target value. Finally, we compute a summary metric. Here we use the F-statistic, which indicates the accuracy of alternative models in correctly classifying each law relative to the baseline.

4.1 Naive Bayes Model 1: Computer Generated Features

The first Naive Bayes Model is based on the document vectors where the data is all the text found in the financial regulatory laws, which includes more than 12,000 distinct words. Each word is a parameter that must be estimated across each of the six classes. We took the raw text of the laws and converted it into document vectors as described in the previous section and estimated the parameters of Naive Bayes Model. This model produced an accuracy of 37% with an F-Measure of 0.38.

Our baseline system is a model that predicts Class 0 for all documents. Absent any other information, the best prediction for a document is a class that has the highest prior probability, which is 0.26 for Class 0. We should note that the Naive Bayes Model 1 based solely on text features does better than the baseline model by 11%.

Table 2 shows the prior probabilities for the six classes of Discretion.

Table 2 Class and prior probability

Class	Prior probability
0	0.26
1	0.14
2	0.25
3	0.24
4	0.08
5	0.07

4.2 Naive Bayes Model 2: Manually Coded Features

We first compare the model with features extracted from the raw text derived from the coding rules outlined above. We take the same set of laws and their corresponding coding rules as features. We identified more than 40 features from the coding rules, including the Number of Provisions with Delegation, constraints such as Reporting Requirements, Time Limits, et cetera. We next created a second Naive Bayes Model using these hand-labeled coding rules as features. Naive Bayes is a general classification algorithm that can take any type of feature vectors as inputs. For Model 2, we again estimated the parameters using the same set of laws that was used to estimate the parameters for building Model 1, and produced an accuracy of 30.0% and F-Measure of 0.40. Interestingly, the raw text model produced a higher level of accuracy than the model built solely from the coding rules. When we build a Naive Bayes Models with manually hand coded features the model parameters are estimated in a similar fashion as stated in Eq. (4) except instead of words w_j we have hand coded features h_k as described in Eq. (6).

$$p(h_1, \ldots, h_m | C_i) = \prod_{k=1}^{m} P(h_k | C_i).$$

(6)

4.3 Naive Bayes Model 3: Combining Manual and Computational Features

Naive Bayes Model 3 combines the purely raw text approach of examining all of the text and the manual approach in which we use the coded features extracted by annotators from the texts. We again estimated the parameters as described in Sect. 4. This model produces an accuracy of 41% and an F-measure of 0.42. These results indicate that a combination of both raw text and manual approaches performs better than either individual approach. When we combine the features we are pooling both sets of w_j and h_k features into same pool. For the estimation of $p(w_i, \ldots, w_n, h_k, \ldots, h_m|C_i)$ we again assume conditional independence among features given the class allowing us to efficiently compute $p(w_i, \ldots, w_n, h_k, \ldots, h_m|C_i)$ using the following equation $\prod_{j=1}^{n} P(w_j|C_i) \cdot \prod_{k=1}^{m} P(h_k|C_i)$.

4.4 Naive Bayes Model 4: Feature Selection Model

The number of parameters for Model 1 is almost the same size as the vocabulary of the corpus, while the total number of parameters for Model 2 equals the number of manually labeled coding rules. It is likely that the raw text-based features can be overwhelming for a small number of manually labeled features. Therefore, we built a fourth Naive Bayes Model where we ran a feature selection algorithm on the combined set of features.

Feature selection algorithms select a subset of features based on different constraints or on the maximization of a given function. We used a correlation-based feature selection algorithm, which selects features that are highly correlated within a given class, but with low correlation across classes, as described in Hall [19]. The feature selection algorithm picked up a feature set containing 47 features, including some features from the manually produced coding rules and a few word-based features as well. Some of the words selected by the feature selection algorithm of Discretion Level include: *auditor, deficit, depository, executives, federal, prohibited, provisions, regulatory, and restrict.*

Model 4 produced the highest level of accuracy at 67% with an F-measure of 0.68. This increase in accuracy is explained in part by the smaller feature set that remains after we discard a number of word-based features. The smaller feature set allows us to better estimate the parameters with our data set of 120 laws thereby reducing the data sparsity problem. The best model produced a high degree of accuracy only after careful feature selection and model design.

4.5 Naive Bayes Model 5: Feature Selection Model with Topics

In the previous model, we combined a selection of word features and three manually coded features. In this last model, we further enrich the analysis by including topics identified by topical modeling as additional features. Topical modeling allows us to validate the model beyond the accuracy measure. Indeed, we argue that different policy sub-domains should have different Discretion Levels, depending on the risk involved for politicians. Thus, we will test whether the topics indicative of more risk indeed predict higher discretion.

Before showing how the topics affect the performance of the Naive Bayes model, we present the results of the topic modeling itself, which shed light on the content of the corpus, and how these results are to be evaluated.

To fit a topic model, one must stipulate the number K of topics (unless we use a Bayesian non-parametric topic model where K is also inferred from data). An analysis typically explores corpora by fitting the topic model with different numbers of topics and examining their relevance for the analysis of interest. Figure 7 presents words representative of the topics for a model where $K = 5$. It shows both the seven words with the highest probability of appearing in a text of a given topic, as well as the seven words that are most representative of each topic according to their high FREX score, a measure that is similar in spirit to TF $-$ IDF, as it combines the exclusivity of a word to a topic and its prevalence.[14]

Through examination of the words, we can interpret the underlying topics inferred by the model. Topic 1 concerns traded securities, including futures and swap agreements. We label it *securities_futures*. Topic 2 concerns the regulation of banks, specifically those laws that seek to insure the safety and soundness of the financial system. We label it *banking_safety*. Topic 3 concerns mortgages and the funding of housing and urban development. We label it *housing*. Topic 4 concerns banking regulation, specifically bank fraud and the relationship of U.S. banks and banking policies to foreign banks and policies. We label it *bank fraud_foreign banks*. Topic 5 concerns the relationship of financial institutions to welfare programs, and market protection. We label it *banking_welfare*. Using our domain expert judgment, these topics appear coherent and meaningful.

With this dataset, we can further evaluate the quality of the topics because we have human-coded information on topics. For each law, annotators determined whether it addresses each of the following six policy issues (and a law can address more than one): banking, securities, commodities, regulation, consumer protection, and mortgage lending. To validate the topics, we regressed each of these annotated labels on $\theta_{1:5}$, that is, the posterior of the topic proportions (for the five latent topics of the topic model). If the latent topics are coherent, these regressions should show thematically logical association between the annotated labels and the latent topics. Table 3 shows that this is indeed the case. For each annotated topic label in the

---

[14]We define the exclusivity score of a word j for a topic k as the ratio of its probability of occurring in topic k to its probability of occurring in other topics. Thus $\phi_{k,j} = \frac{\beta_{k,j}}{\sum_{i \neq k} \beta_{ij}}$. We then define the FREX$_{k,j}$ score as the harmonic mean of the word's rank in the distribution of exclusivity scores for topic k (which frequency distribution is denoted $\phi_{k,\cdot}$) and the word's rank in the distribution of word frequencies for topic k (which frequency distribution is denoted $\mu_{k,\cdot}$). Thus:

$$\text{FREX}_{k,j} = \left(\frac{\omega}{\text{ECDF}_{\phi_{k,\cdot}}(\phi_{k,j})} + \frac{(1 - \omega)}{\text{ECDF}_{\mu_{k,\cdot}}(\mu_{k,j})} \right)^{-1} \tag{7}$$

where ω is the weight for the exclusivity (which is set to 0.5 by default) and ECDF$_{x_{k,\cdot}}$ is the empirical cumulative density function applied to the values x over the first index, giving us the rank. See Airoldi at al. [1], p. 280.

Topic 1: securities-futures	Topic 2: banking-safety	Topic 3: housing
Gramm-Leach-Bliley	Sarbanes	FHA-insur
self-regulatory	Glass-Steagal	handicap
NRSRO	SAIF	FNMA
SIPC	CRA	homeownership
Security-based	Riegle	HUD
swap	FSLIC	rent
CFTC	Greenspan	FHA
exchange	RTC	urban
commission	thrift	flood
trade	FDIC	mortgage
futures	union	charge
commodity	deposit	area
transaction	save	community
securities	consumer	grant

Topic 4: bank fraud - foreign banks	Topic 5: welfare-banking
anticrim	medicaid
FRB	medicar
balance-of-payment	FY
habea	food
Patman	crop
gun	health
death	secretariat
merger	benefit
foreign	appropriation
US	payment
tax	assistance
holding	security
court	tax
percent	grant

Fig. 7 Topic model for the 120 laws with 5 topics. Each topic is summarized by its seven most frequent words and its seven words with the highest FREX score. Larger words are more frequent. Each word's position on the x-axis indicates its exclusivity to the topic.

Topic 1 terms and acronyms—Gramm-Leach-Bliley: a 1999 law that deregulated financial markets. NRSRO: Nationally Recognized Statistical Rating Organization, credit rating agency. SIPC: Securities Investor Protection Corporation. CFTC: Commodity Futures Trading Corporation, an agency that regulates futures and options markets.

Topic 2 terms and acronyms—RTC: Resolution Trust Corporation, a federal agency that operated between 1989 and 1996 and administered insolvent federal savings and loan institutions. FDIC: Federal Deposit Insurance Corporation, an agency which provides deposit insurance for banks. FED: Federal Reserve Bank. FSLIC: Federal Savings and Loan Insurance Corporation, an agency that until 1989 administered the insurance of federal savings and loan institutions. CRA: Community Reinvestment Act, a law to incentivize banks to meet the financial needs of lower income communities, particularly regarding mortgage lending. SAIF: Savings Association Insurance Fund.

Topic 3 terms and acronyms—HUD: U.S. Department of Housing and Urban Development. FNMA: Federal National Mortgage Association (Fannie Mae), providing mortgage-backed securities. FHA: Federal Housing Administration, a federal agency setting construction standards and insuring loans for home building.

Topic 4 terms: Patman refers to Wright Patman who was the chair of the Senate Committee on Banking and Currency for 10 years.

Topic 5 terms and acronyms—FY: fiscal year, a term used in appropriations bill.

Table 3 Table showing the association between annotated labels (human coded topics on the left) and the latent topics identified by the topic model

Human annotated topic labels	Topics identified by the topic model	
	Positive association	Negative association
Banking	Banking_safety	Securities_futures
	Bank fraud_foreign banks	
	Welfare_banking	
Regulation	Banking_safety *interacted with*	Securities_futures
	bank fraud_foreign banks	
Consumer protection	Housing	Securities-futures
	Banking_security	Bank fraud_foreign banks
Commodities	Welfare_banking	Banking_safety
		Housing
		Bank fraud_foreign banks
Securities	Securities_futures	banking_safety
		housing
Mortgage_lending	Housing	Banking_safety
	Welfare_banking	Securities_futures
		Bank fraud_foreign banks

original dataset, the table shows which latent topics identified by the topic model are significantly associated with it (either positively or negatively). For example, we see that laws with annotated label *banking* are strongly associated with latent topics *banking-security, bank fraud—foreign banks, welfare-banking*, but negatively associated with *security*. Conversely, the annotated label *securities* is positively associated with the topic *securities-futures*, but negatively associated with *banking-safety* and *housing*. We see from these results that the topics identified by the topic model are coherent and reliable indicators of existing policy sub-domains.

How do we know that $K = 5$ is an appropriate number of topics? There is no "best" number of topics, since the appropriate resolution depends on the interpretation and insights sought by the analyst. However, K could be too small (lumping topics that are quite different) or too large (splintering the data into groups that are difficult to interpret because not truly distinguishable). When the number of topics is small, these problems appear by inspection (as we have done above), but when there are many topics, quantitative indicators of the quality of the model are useful. We, therefore, quickly present a few below.

The appropriate methods of validating a topic model is an active topic of research [6]. One approach is to compute the probability of held-out data. Figure 8a shows the perplexity of held-out data in tenfold cross-validations for models with different number of topics K. Perplexity is $p = -\exp(\frac{\sum_{i \in \text{heldout set}} p(y_i | \beta_{1:K}, \alpha)}{\text{total \# words in heldout set}})$. A lower perplexity indicates a better fit. For our data, the perplexity declines rapidly as a function of K while $K \leq 4$, after which increases in K have a diminishing impact on this measure of fit.

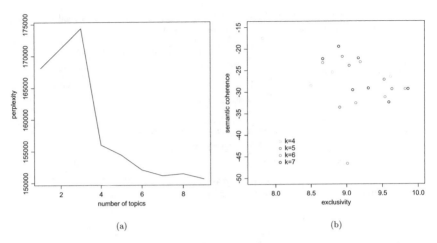

(a)   (b)

Fig. 8 Different measures for comparing topic models (**a**) Perplexity as a function of K (**b**) Exclusivity and semantic coherence for different values of K

Recent research demonstrates that probability-based measures of fit, such as perplexity, correlate poorly with human judgment of the coherence and interpretability of topics [9]. In complement to probability-based measures, researchers have proposed measures that capture the goals of the model, namely finding coherent and exclusive topics. Mimno et al. [29] introduced a measure of *semantic coherence* of topics. Semantic coherence computes how frequently the most probable words of topic k co-occur in the same document (relative to the baseline frequency of these words). This is an intuitive measure of how clustered the words representing topic k are in the corpus. However, semantic coherence will decrease with the number of topics (since the most frequent words in each topic become more likely to appear in multiple topics). It must be traded against *exclusivity*, a measure of how exclusive the words of a topic are to that particular topic.[15] Figure 8b plots the exclusivity and the semantic coherence of each topic as we vary K from 4 to 7. Intuitively, models whose topics are on the frontier of exclusivity versus semantic coherence are of better quality [37]. In our case, we see that $K = 4$ and $K = 6$ yield each one topic that scores low either on semantic coherence or on exclusivity. Topics obtained with $K = 5$ and $K = 7$ are closer to the frontier. In fact, $K = 7$ seems slightly superior. Examining the resulting topics, we observed that topics 1 and 5 are split in finer categories that are easy to interpret in the context of financial regulation. However, for reasons of parsimony given the size of the corpus, we keep $K = 5$ in the experiments of Sect. 4.

---

[15]Exclusivity of word j to topic k was defined in an earlier footnote, and is $\phi_{k,j} = \beta_{k,j} / \sum_{i \neq k} \beta_{i,j}$. The exclusivity score for the whole topic is the sum of these $\phi_{k,j}$ word scores for all words in a topic.

We now turn to the final predictive model, in which the topics of the topic model are added to the previously selected features. We wish to know whether the policy sub-domains are useful in predicting the discretion level. To answer this question, we associate each law with its dominant topic (the topic k for which $\theta_{i,k}$ is highest), as derived above, and add this topic as an attribute in the Naive Bayes Feature Selection Model 4, for a total of 48 features.

Adding topics as additional features increased the classification accuracy of the model to 70.83%. This model produces an F-measure of 0.71. Furthermore, a chi-square test rejects the null hypothesis of independence between the discretion classes and the topics. These results reinforce our notion that combining different types of features generated through alternative methods strengthens the quality of the model. Here we add the latent variables inferred by a topic model, corresponding to different policy domains within financial regulation legislation.

As mentioned earlier, we can go further and apply these topics as a test that the model is using these features as predicted by theory. Coming back to the political economy hypotheses outlined in the first part of the paper, we expect that politicians will grant less discretion in policy areas that entail more political risk. Topics 1 and 2, which concern securities, consumer protection and the stability of the financial system are high risk (coded "H"). Topics 3 and 5, concerning the financing of homes and welfare programs is medium risk (coded "M"), while topic 4, the regulation of bank fraud and foreign banks is low political risk (coded "L"). If we use this risk level as an attribute instead of the topic (keeping our number of attributes at 48), we recover the identical classification accuracy of 70.83. This indicates that the risk levels are meaningful predictors.

We run an ordinal logistic regression of the discretion level on the risk level of the law's policy domain to analyze the influence of the risk level of the policy domain on the discretion level. For each discretion level C_i, Fig. 9 shows the first difference between being a law in a High Risk policy domain versus a Low Risk policy domain. This quantity indicates the changes in the expected probability of the law having level C_i when the we shift the variable "political risk" from low to high. For example, for discretion level 0 ($C_i = 0$ the lowest discretion level) the figure represents the difference in the expected probability that a law has discretion level 0 given that it addresses a High Risk policy domain versus a Low Risk policy domain: $E(C_i = 0|H) - E(C_i = 0|L)$. When Congress and the executive disagree over policy saliency, we see from Fig. 9 that laws concerning High Risk policy domains

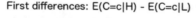

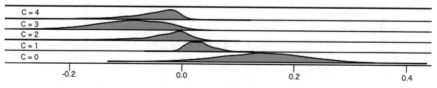

Fig. 9 Probability distribution of the first difference of falling in each discretion class between a law covering a high risk policy domain versus a low risk policy domain

are less likely to have high levels of discretion, in line with our hypothesis. This demonstrates that the features affect the prediction of discretion in a way that is consistent with theory.

5 Discussion and Conclusions

The results of the five models are summarized in Table 4. Model 5 performs best, yielding a 71% accuracy level. It includes three types of features: computationally selected features, manually coded features, and topics, all which complement each other by capturing different attributes embedded in the financial regulation laws.

Let us consider how the computational techniques described here would concretely modify the research design of an observational study in political economy. First, we see that the Naive Bayes Model allows us to quantify the measurement uncertainty inherent in mapping complex texts onto an ordinal or categorical variable. This uncertainty can in turn be integrated in the statistical analysis of the data to better quantify our uncertainty. Hence, these methods can be useful even with a small hand-annotated corpus, simply to put errors bars on the coding of the data.

Second, by quantifying how the measurement uncertainty resulting from the coding changes as we modify the type of features used, the methods allow us to decide how to spend scarce resources to develop our datasets. In the example used in this paper, the hand-annotated *Discretion Index* (in Sect. 2.2) is very time consuming and prone to errors as each provision of each text needs to be carefully annotated for whether it delegates authority and the number of procedural constraints. This approach is not scalable. In contrast, Models 1, 2, and 3 are perfectly scalable, as they require no human coding. Yet, they lead to high levels of uncertainty. Models 4 and 5 include manually coded features which are much less time consuming and less prone to error than the Discretion Index and they help reduce the uncertainty. We can include these different uncertainty levels in power calculations and decide the degree to which it is best to increase the quality of the coding by including more or less costly human-annotated features or increase the quantity of data by eschewing the more costly human-annotated features and relying more heavily on computationally generated features.

Table 4 Naive Bayes models

Feature Type	Accuracy(%)	*F*-Measure
Model 1: computer generated text features (C)	36.66	0.38
Model 2: manual-coded variables/features (M)	30.00	0.40
Model 3: C + M	40.83	0.42
Model 4: feature selection (C + M)	66.66	0.68
Model 5: feature selection with topics (C + M + T)	70.83	0.71

This paper is the first to our knowledge to apply this combination of supervised and unsupervised models to legal texts for the purpose of testing theories in political economy. We have focused on the simplest text data representation, the bag-of-words approach. We have found that this data representation and topic modeling successfully identify coherent policy sub-domains. We have also found that these word features help improve Naive Bayes classifiers in predicting institutional variable, such as regulatory discretion. While the level of accuracy attained is sufficient for some analyses,[16] it is only 71% in our best model. This indicates that a bag-of-words approach has limits.

Subsequent research to enhance the methods used in this paper include feature-level improvements, such as N-grams with high term-hood (e.g., Wong [44]); transforming the word features and the texts into synsets, with Wordnet; and identifying features based on Neural Nets and Deep Nets (e.g., Mikolov et al. [30]). A second class of enhancements include improving the algorithm by adopting random forests, support-vector machines or Maximum Entropy Classifiers that use features combining words, semantic labels and part-of-speech-tags [33].

As we know, computational data science captures complex patterns and interactions that are not easily recognized by manual coding rules. However, these NLP and ML techniques have not been used to analyze text-based data to test theories of regulatory design. These computational methods enable us to represent the text of a given law as a feature vector, where each feature represents a word or weighted terms for words. We also apply topic modeling to associate topics with institutional features, such as regulatory discretion. Each of these techniques provides potential improvements over manual-coding and inferences from a set of defined coding rules. Yet these computational models rely on the critical data initially produced by subject matter experts to inform or "seed" the model and train these complex algorithms. Computational data science techniques, therefore, are an important and critical complement to observational studies.

A research strategy that uses more than one technique of data collection and interpretation can improve the validity of analyzing high-dimensional datasets commonly found in political economy studies of financial regulation. The practical implications of the analysis are manifold. The analytical methods developed enable governments and financial market participants alike to: (1) automatically score policy choices and link them to various indicators of financial sector performance; (2) simulate the impact of various policies or combinations of policy under varying economic and political conditions; and (3) detect the rate of change of market innovation by comparing trends of policy efficacy over time. The analysis will help governments to better evaluate the effect of the policy choices they confront, as well as assist the financial community to better understand the impact of those choices on the competitive environments they face.

---

[16]For a test of these hypotheses using standard regression analysis, see Groll et al. [16].

References

1. Airoldi EM, Blei DM, Erosheva EA, Fienberg SE. Handbook of mixed membership models and their applications. Chapman and Hall/CRC handbooks of modern statistical methods (Page A). Boca Raton: CRC Press; 2015 (Kindle Edition).
2. Alesina A, Tabellini G. Bureaucrats or politicians? Part I: a single policy task. Am Econ Rev. 2007;97(1):169–79.
3. Alonso R, Matouschek N. Optimal delegation. Rev Econ Stud. 2008;75(1):259–93.
4. Barth JR, Caprio G Jr., Levine R. Rethinking banking regulation: till angels govern. New York: Cambridge University Press; 2006.
5. Bendor J, Meirowitz A. Spatial models of delegation. Am Polit Sci Rev. 2004;98(2):293–310.
6. Blei DM. Probabilistic topic models. Commun ACM. 2012;55(4):77–84.
7. Blei DM, Ng AY, Jordan M. Latent Dirichlet allocation. J Mach Learn Res. 2003;3:993–1022.
8. Brill E. A simple rule-based part of speech tagger. In: Proceedings of the workshop on speech and natural language. Association for computational linguistics; 1992. p. 112–116.
9. Chang J, Boyd-Graber J, Wang C, Gerrish S, Blei DM. Reading tea leaves: How humans interpret topic models. Adv Neural Inf Proces Syst. 2009;(22):288–96.
10. Epstein D, O'Halloran S. Delegating powers New York: Cambridge University Press; 1999.
11. Gailmard S. Discretion rather than rules: choice of instruments to constrain bureaucratic policy-making. Polit Anal. 2009;17(1):25–44.
12. Gailmard S, Patty JW. Slackers and Zealots: civil service, policy discretion, and bureaucratic expertise. Am J Polit Sci. 2007;51(4):873–89.
13. Gailmard S, Patty J. Formal models of bureaucracy. Ann Rev Polit Sci. 2012;15:353–77.
14. Griffiths TL, Steyvers M. Finding scientific topics. Proc Natl Acad Sci. 2004;101(suppl 1):5228–35.
15. Griffiths TL, Steyvers M, Tenenbaum JB. Topics in semantic representation. Psychol Rev. 2007;114(2):211–44.
16. Groll T, O'Halloran S, McAllister G. Delegation and the regulation of financial markets, mimeo; 2015.
17. Grün B, Hornik K. Topicmodels: An R Package for Fitting Topic Models. Journal of Statistical Software, 2011;40(13):1–30.
18. Haber S. Political Institutions and Financial development: evidence from the political economy of bank regulation in Mexico and the United States. In: Haber N, Weingast, editors. Political Institutions and Financial Development. Stanford: Stanford University Press; 2008
19. Hall MA. Correlation-based feature subset selection for machine learning. Phd Thesis. University of Waikato; 1998.
20. Keefer P. Beyond legal origin and checks and balances: political credibility, citizen information, and financial sector development. In: Haber N, Weingast, editors. Political Institutions and Financial Development. Stanford: Stanford University Press; 2008.
21. Kroszner R, Strahan P. What drives deregulation? economics and politics of the relaxation of bank branching restrictions. Q J Econ. 1999;114(4):1437–67.
22. Maskin E, Tirole J. The politician and the judge: accountability in Government. Am Econ Rev. 2004;94(4); 1034–54.
23. McCallum A, Nigam K. A comparison of event models for naive Bayes text classification. In AAAI-98 workshop on learning for text categorization, vol. 752. 1998. p. 41–8.
24. McCubbins MD, Schwartz T. Congressional oversight overlooked: police patrols versus fire alarms. Am J Polit Sci. 1984;28(1):165–79.
25. McCubbins MD, Noll R, Weingast B. Administrative procedures as instruments of political control. J Law Econ Org. 1987;3;243–77.
26. McCubbins MD, Noll R, Weingast B. Structure and process, politics and policy: administrative arrangements and the political control of agencies. Virginia Law Rev. 1989;75:431–82
27. Melumad ND, Shibano T. Communication in settings with no transfers. RAND J Econ. 1991;22(2):173–98

28. Mihalcea R, Radev D. Graph-based natural language processing and information retrieval. Cambridge; Cambridge University Press; 2011
29. Mimno D, Wallach HM, Talley E, Leenders M, McCallum A. Optimizing semantic coherence in topic models. Proceedings of the conference on empirical methods in natural language processing. Association for Computational Linguistics; 2011
30. Mikolov T, Ilya S, Kai C, Corrado GS, Dean J. Distributed representations of words and phrases and their compositionality. Advances in neural information processing systems; 2013.
31. Morgan DP. Rating banks: risk and uncertainty in an opaque industry. Am Econ Rev. 2002;92(4):874–88.
32. Nadeau D, Sekine S. A survey of named entity recognition and classification. Lingvisticae Investigationes 2007;30(1):3–26.
33. Nigam K, Lafferty J, McCallum, A. 1999. Using maximum entropy for text classification. IJCAI-99 Workshop on machine learning for information filtering.
34. O'Halloran, S. Politics, process, and American trade policy. Ann Arbor: University of Michigan Press; 1994
35. O'Halloran S, Maskey S, McAllister G, Park DK, Cheng K. Data science and political economy: application to financial regulatory structure. J Soc. Sci. 2016;2(7):87–109.
36. Philippon T, Reshef A. Wages and human capital in the U.S. financial industry: 1909–2006. Q J Econ. 2012;112(4):1551–1609.
37. Roberts ME, Stewart BM, Tingley D. STM: R package for structural topic models. R package. 2014 Jun;1:12.
38. Roberts ME, Steward BM, Tingley D, Lucas C, Leder-Luis J, Gadarian SK, et al. Structural topic models for open ended survey responses. Am J Polit Sci. 2014;58(4):1064–82.
39. Spärck Jones K. A statistical interpretation of term specificity and its application in retrieval. J Doc. 1972;28:11–21.
40. Steinbach M, Karypis G, Kumar V. A comparison of document clustering techniques. In: Proceedings of the 6th ACM SIGKDD, World text mining conference, Boston, MA; 2000
41. Volden C. A formal model of the politics of delegation in a separation of powers system. Am J Polit Sci. 2002;46(1):111–33.
42. Volden C, Wiseman A. formal approaches to the study of congress. In: Schickler E, Lee F, editors. Oxford handbook of congress. Oxford: Oxford University Press; 2011. p. 36–65.
43. Wiseman, AE. Delegation and positive-sum bureaucracies. J Polit. 2009;71(3):998–1014.
44. Wong W, Liu W, Bennamoun M. Determination of unithood and termhood for term recognition. Handbook of research on text and web mining technologies. Hershey; IGI Global; 2008.
45. Wooldridge, Jeffrey, Introductory Econometrics Paperback, South-Western College Publishing, 2006.

Frequent and Non-frequent Sequential Itemsets Detection

Konstantinos F. Xylogiannopoulos, Panagiotis Karampelas, and Reda Alhajj

1 Introduction

Detecting frequent itemsets is one of the most important problems of data mining addressed firstly by Agrawal and Srikant [1]. Sequential frequent itemsets detection as introduced in [1] is a special case of the general problem of detecting sequential itemsets and due to many difficulties it presents, many scientists have turned their attention to it. It has many fields of implementation in business, marketing, finance, insurance, etc. For example, sequential frequent itemsets detection can be used when we want to examine how frequently "a new car is bought and a new insurance policy is bought immediately after the car." A more representative example of sequential itemsets (or patterns) detection can be derived from car industry as follows: Whenever a car is sold, the owner has to go through a service process which can include, e.g., engine oil change, tires change, brakes fluid change, service after 40,000 km, etc. In this sequence of events an engine or any other part of the car may fail. The car industry is very interested in detecting when a failure occurs in the sequence of service events, in order to reform the timeline of services or

K.F. Xylogiannopoulos (✉)
Department of Computer Science, University of Calgary, Calgary, AB, Canada
e-mail: kostasfx@yahoo.gr; kostasfx@outlook.com

P. Karampelas
Department of Informatics and Computers, Hellenic Air Force Academy, Dekelia Air Base, Attica, Greece
e-mail: pkarampelas@gmail.com

R. Alhajj
Department of Computer Science, University of Calgary, Calgary, AB, Canada

Department of Computer Science, Global University, Beirut, Lebanon
e-mail: alhajj@ucalgary.ca

© Springer International Publishing AG 2017
M. Kaya et al. (eds.), *From Social Data Mining and Analysis to Prediction and Community Detection*, Lecture Notes in Social Networks,
DOI 10.1007/978-3-319-51367-6_9

make specific parts more durable, etc. In the above-mentioned example the order of events is essential. It is completely different to say "oil change, brakes fluid change" than "breaks fluid change, oil change." That is an illustrative definition of sequential itemsets, assuming that each event is an item. Moreover, each event in the above-mentioned example, or generally speaking each item, can reoccur, such as oil change.

The problem of sequential frequent itemsets detection is also very similar to the problem of detecting repeated patterns in strings. In such a case, we have a very long string of discreet values based on a predefined alphabet and we want to detect patterns that exist in the string (e.g., in DNA). Yet, in the general case of sequential frequent itemsets detection, we have transactions constructed from items and we want to detect sequential itemsets that exist. Common methods to address the specific problem are to either use apriori type or pattern growth type algorithms. Lately, hybrid and early pruning algorithms have also been presented that can address the specific problem adequately.

In the current paper, we will address the general problem as we will define it here and not as it is defined in [1, 2] by Agrawal and Srikant. More particularly, in [1] an itemset is defined as "a non-empty set of items." (p. 3). Yet, it is important to mention that the specific definition does not conform with the strict mathematical definition of set, as it derives from Set Theory. As we can read in [3] "a set is formally defined to be a collection of distinct elements." Generally, a set is a collection (or group) of distinct and strictly defined objects of thought or perception conceived as a whole, where each object is called element or member of the set, according to Cantor's definition [4]. On the contrary, a sequence is an ordered collection of objects for which reoccurrence is allowed. Since in data mining, for pattern detection purposes, we care about sequences where their objects can reoccur, we can also accept this for collections of items by strictly state the deviation from the mathematical definition. We can allow having an item to occur more than once in an itemset, e.g., the DNA sequence can also be considered a set of proteins constructed from alphabet {A, C, G, T} where every arrangement of proteins (pattern) can also reoccur in the DNA sequence. Therefore, in order to be accurate we have to redefine itemset, for data mining purposes, as a collection of items where the reoccurrence of each item is allowed, while the detection of sequential itemsets can be defined as the detection of ordered itemsets in a sequence, for which the repetition of an item is allowed.

Therefore, we can redefine the sequential itemsets detection problem as the detection of itemsets (as defined above) which can occur either in the same sequence (also expressed as transaction) and/or in different sequences in a collection of sequences. The above definition is the generic description of the problem because we cover every possible combination of itemsets detection, i.e., single item itemsets (e.g., having only one element, where of course sequence cannot be directly defined), itemsets containing a whole sequence of items, itemsets that appear only in different sequences, itemsets that appear multiple times in the same sequence, and finally itemsets that occur either in the same sequence and in different sequences.

In order to address the problem of detecting sequential frequent itemsets, we will use a variation of our methodology presented in [5, 6] for detecting all repeated patterns in a string. The proposed methodology will allow us to detect not only the most frequent sequential itemsets but also all itemsets that exist at least twice in a database of transactions. We will show that the newly proposed methodology, Sequential All Frequent Itemsets Detection (SAFID), is very efficient and can be used in any transactions' database with very good results. Detailed experimental analysis will also prove the efficiency and novelty of the new methodology.

The novelty of the proposed methodology is based on two factors: (a) it can detect rapidly every sequential itemset and (b) the detection can be done regardless of any kind of support threshold. As we have shown in [6] the detection of every repeated pattern can take on average less than 0.2 ms and therefore even when thousands of patterns have to be detected the overall time is not a significant problem. Moreover, although the detection of every sequential itemset, and therefore itemsets of very low frequency, seems not significant, in some cases this can be proven to be very useful especially when outliers are detected, e.g., in case of fraud detection, anti-terrorism surveillance and analysis, etc. Furthermore, by performing the specific analysis periodically we can detect trends in low frequency itemsets which may increase/decrease slower or faster and thus lead us to proceed to appropriate actions, e.g., in developing different marketing policies and strategies. Alternatively, the proposed methodology can also be used to specifically detect low frequency itemsets, something that, to the best of our knowledge, does not exist in the literature.

The rest of the paper is organized as follows: Sect. 2 reviews the related work. Section 3 describes the proposed methodology. Section 4 discusses the experimental analysis. Finally, Sect. 5 presents the conclusions and anticipated future work.

2 Related Work

Sequential pattern mining has occupied the data mining research community from the 1990s where the first attempts were made to address the problem of identifying itemsets that appear together in transactions [1]. The specific problem rapidly was broadened in various applications domains such as healthcare to detect common patterns detected in diseases, education to assess teamwork collaboration in software development projects, web usage mining to identify what pages have been seen together by the users with the purpose to understand the navigational patterns of the visitors [7], text mining to discover trends in blogs or for documents classification, bioinformatics, e.g., for protein fold recognition, telecommunications, e.g., to predict the movement of a mobile user, and intrusion detection to detect common patterns in Distributed Denial of Service attacks [8].

According to [9], sequential pattern mining algorithms are distinguished in three broad categories: apriori type algorithms, pattern growth algorithms, and hybrid methods. In a more recent review of the techniques used in sequential pattern mining [7] the categories of techniques identified are four, with the addition of the early pruning methods. Further taxonomy can be applied inside the broad categories based on the organization or on the type of information of the datasets mined.

In this regard, when the data are organized horizontally including the identification information and time relevant fields, there are several algorithms in the literature that can be applied such as Generalized Sequential Patterns (GSP) [2], PSP [10], SPIRIT [11], MFS [12], and MSPS [13]. GSP algorithm [2], in order to detect the frequent itemsets, makes multiple passes over the dataset and detects candidate itemsets and their corresponding support in each pass until there is no frequent or candidate sequences. PSP algorithm [10] is based on GSP and uses an intermediary data structure to store the candidate and frequent itemsets that improve the efficiency of GSP. The SPIRIT family algorithms [11] introduce regular expressions as constraints to the mining process. MFS algorithm [12] improves GSP process by mining a sample from the dataset to estimate the candidate itemsets and then, based on these, the frequent itemsets are detected. A similar sampling technique is used in MSPS algorithm [13] to reduce the steps required for completing the mining process.

When the items are associated sequentially with events then it is possible to apply depth-first algorithms to detect the sequence of the itemsets. Such algorithms are SPADE [14], SPAM [15], LAPIN [16], CCSM [17], IBM [18]. SPADE algorithm [14] adopts a different approach versus the previously mentioned algorithms by re-organizing the dataset using vertical id-lists and decomposing the original dataset in smaller subsets based on some combinatorial properties. SPAM algorithm [15] uses a vertical bitmap representation and a depth-first traversal in order to identify all the frequent patterns. LAPIN family of algorithms [16] is based on a vertical id-list that reduces the number of scans required to identify the frequent itemsets. In [7] LAPIN algorithms are considered to belong in the early pruning category since these algorithms use the last position of an item in the sequence early in the data mining process to identify whether an item can be appended to a given sequence and provide a candidate frequent sequence or eliminate the sequence. CCSM algorithm [17] uses a similar id-list to identify the candidate itemsets using k-way intersections to identify the candidate itemsets that are stored in a cache for future use. Finally, IBM algorithm [18] adopts a bit map representation using an indexed structure for mining the frequent itemsets in one pass. In the second broad category, the pattern growth algorithms, the database is split into projected itemsets databases that can then be mined separately [9]. Such type of algorithms based on FP-GROWTH are FREESPAN [19], PREFIX [20], LPMINER [21]. More specifically, FREESPAN algorithm [19] adopts this approach while PREFIX algorithm [20] uses prefix projection in order to reduce the size of the sequence database and as a result the generation of candidate frequent itemsets. LPMINER algorithm [21] introduces the smallest valid extension property to reduce the required space to search for frequent itemsets. There are more hybrid algorithms that attempt to improve performance by combining techniques and features of both of the aforementioned categories of algorithms such as ApproxMAP [22] which mines the approximate frequent itemsets since as it is argued in [22] all the frequent itemsets may not be of interest while they require long time processing. According to [7] other early-pruning algorithms include the first-Horizontal-last-Vertical Scanning database Mining algorithm (HVSM) [23] which is based on

SPAM with several improvements including the early pruning of large itemsets that do not appear in subsequent scans taking sibling-nodes as children-nodes and thus improving the overall efficiency of the algorithm. Another algorithm that belongs to earlypruning category is the Direct Sequence Comparison (DISC) [24] which prunes early in the process the sequences that are infrequent comparing them with the other of the same length. By doing this, DISC algorithm is able to perform faster than other algorithms since it is not necessary to count the support for the infrequent sequences.

In recent literature [25, 26], more complex algorithms have appeared that apart from the features that the traditional frequent sequential pattern mining techniques are taking into account, e.g., number of items, they use additional attributes such as the quantity of a specific item or even the price or profit. These algorithms are considered to belong in an emerging mining category called High Utility Sequential Patterns (HUSP) mining techniques. Such algorithms are the USpan [25] which utilizes a lexicographic sequence tree with the utility, i.e., profit from a specific item, two concatenation mechanisms that are used to select the items that will be used to create the descendants and a pruning technique utilized to select which items will be used in concatenation. The experiments presented in [25] show that the algorithm is relatively efficient in calculating high utility sequences. Another algorithm in the same category is the HUSP-Stream [26] which introduces a new data structure HUSP-tree that maintains the necessary information for the utility of the sequences and runs in three phases to mine the high utility sequential data.

The methodology proposed in this paper, as stated at the beginning, utilizes a variation of our previous work [5] for detecting all repeated patterns in a string. A similar methodology that claims that finds maximal non-gapped motifs or in other words maximal and minimal repeated patterns in a string is presented by Ukkonen in [27]. This methodology potentially could be altered to mine sequential frequent maximal and minimal itemsets as a special case of the general problem as it has been stated here. While in [27] it is proved that this can be achieved in $O(n)$ there are no experimental results to apply the theoretical method and as a result our method cannot be directly compared with that proposed in [27] even for this special case. Our proposed methodology uses an aggressive approach by finding all the sequential frequent sequences in $O(n \log n)$ complexity that outperforms all the known implemented algorithms of any type, to the best of our knowledge.

3 Proposed Methodology

In the current paper, we present a novel methodology based on the ARPaD algorithm presented in [5] which allows the detection of all repeated patterns in a string. ARPaD algorithm has been extensively discussed in [5] and proven to be correct, finite (terminates), and has worst case complexity log-linear $O(n \log n)$. Moreover, the specific algorithm has been proven to be extremely efficient and fast since we have managed to analyze strings from 100 million characters length [28] up to

68 billion characters length [6] in a logical time span (few hours) using typical desktop or laptop hardware configuration. On average the algorithm needed for the specific experiments 0.17 ms and despite the billions of patterns detected (approximately 25 billion) the overall time was only 51 days [6]. Although it may seem strange how these numbers can be compared to itemset detection in transactions, it is important to mention that strings' lengths represent also number of records, which in this case can represent transactions. In order this to be achieved a new data structure has been introduced in [6], the Longest Expected Repeated Pattern-Reduced Suffix Array (LERP-RSA), which is a variation of the suffix array and can be defined as (p. 20).

Definition (LERP Reduced Suffix Array) We define as LERP Reduced Suffix Array (LERP-RSA) of a string S, constructed from a finite alphabet Σ, the array of the actual lexicographically sorted suffix strings such that their lengths have been upper-bounded to the length of the LERP. Any suffix string with length larger than LERP has been truncated (at the end) to a length equal to LERP. The LERP-RSA has one column for the index of the position where each suffix string occurs in the string and another column for the reduced suffix strings.

The specific data structure has many important advantages comparing to any data structure used for pattern detection. Major advantages are the classification of the data structure to many and much smaller classes, which can be easily stored to any local or distributed database management system or file system, and the ability of parallel processing of the data structure through the ARPaD parallel execution. This means that the total time needed for the analysis and detection of all repeated patterns can be significantly reduced when the appropriate resources are available, i.e., normal PCs with multiple cores. This reduction, according to what it has been proven in [6], can be expressed as power fractions of the alphabet size, i.e., for an alphabet of size ten (e.g., decimal system) it can be 10, 100, 1000 or even more times faster, based on the classification level it has been used [6]. Therefore, if, for example, the average execution time for pattern detection over sequences constructed from the decimal system $\{0, 1, 2, 3, 4, 5, 6, 7, 8, 9\}$ is t for single class execution, then for Classification Level 1 is $\frac{t}{10}$ for Classification Level 2 is $\frac{t}{100}$, etc. [6].

Although the above briefly described methodology seems irrelevant to the problem of the detection of sequential frequent itemsets, we will present here how this can help us to detect all sequential itemsets that exist in a database, regardless of the frequency of occurrences. The LERP-RSA data structure is a variation of the suffix array data structure, which is used for the analysis of strings. First all suffix strings are created from the original string and then they are lexicographically sorted. After the data structure creation phase ARPaD algorithm runs and detects all repeated patterns that exist at least twice in the string. If we consider each transaction as a string, then we can follow the same process and create the suffix array for the specific transaction. Of course we cannot detect in just one transaction any repeated pattern (itemset in our case) because it simply does not exist. In order to have any kind of repetition in a single transaction we must have repeated items. Usually this is not the case despite the fact that there are cases when this may occur, e.g., when we recording webpage visits by users when in a single session a user may go back and forth and repeat in some cases the same webpage visiting process on a website. In this example, ARPaD algorithm will detect these repeated patterns (itemsets). However, in the general case we care to find frequent itemsets among different transactions. In this case we can create a suffix array for each transaction and then combine all suffix arrays together. Then ARPaD algorithm can detect all repeated patterns (itemsets) among all transaction, which have been inserted in the database.

Usually all other methods for sequential frequent itemsets detection follow a two phase process. First, the transactions are entered in the database and then an algorithm scans the database to detect the sequential frequent itemsets. For this purpose, a support level (i.e., minimum frequency) is used since these methods care to detect itemsets that occur with high frequency in the database. Another reason for this is that no method can handle the enormous number of extremely small itemsets may occur in the database (in strings of length 1 billion digits we have found more than 400 million). The data structure behind such methods (arrays or trees) cannot scale up when the number of transactions and items is big and no algorithm can detect every itemset in a feasible amount of time. Usually, for just one million transactions, current methods may need many GB of available RAM in order to create the data structure. In our methodology, support is not an important parameter and can be omitted since ARPaD algorithm can detect all repeated patterns that occur at least twice and in a very efficient amount of time and by using limited hardware resources, as presented in [5, 6, 28]. In such a case, the algorithm can detect even an itemset that occurs only 2 times in a million transactions which means that has a support of only 0.0002%. Although this may not be an important itemset when executing frequent itemsets detection process, however, as mentioned in the introduction, these low frequency itemsets could be potentially important for specific uses such as anti-frauds, anti-terrorism, and marketing.

The proposed methodology has been divided into five different steps by creating a two phase's process as shown in "Fig. 1." The first phase is the pre-process analysis and the second phase is the ARPaD data mining process. The first phase has three steps and it is important for the transactions' transformation to strings and pre-statistical analysis of very important variables. The second phase is consisted of two steps which is the actual data mining process using the ARPaD algorithm in order to detect all repeated patterns and therefore every frequent and non-frequent itemset and any meta-analyses processes. The complete method can be described as follows.

3.1 Pre-process Analysis Phase

In the Pre-process Analysis Phase, it is important to convert the transaction to string and perform a statistical analysis to some very important characteristics of the transaction, i.e., count the number of items in the transaction and classify all items based on the alphabet used to symbolize them. For example, if items are numbered

Fig. 1 SAFID flow diagram

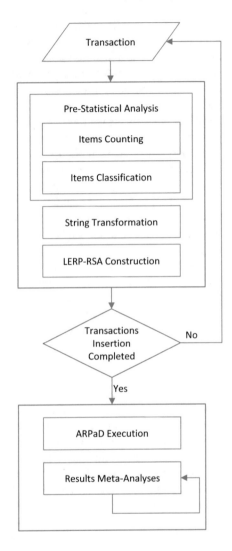

using the decimal system, then the alphabet is {0, 1, 2, 3, 4, 5, 6, 7, 8, 9} while if only letters are used to code items, then the alphabet can be the English alphabet or any other. Usually we use digits or combination of digits and characters from the Latin alphabet because each store or product catalog has specific coding for items known as Stock Keeping Unit (SKU), which can be used directly instead of creating a special dictionary in order to assign new coding to each item. The last step of the first phase is the construction of the LERP-RSA data structure for each transaction (i.e., string) that imports the database. This step can also be executed for all strings separately when all transactions have been imported to the database. Yet, it is much faster to perform this step immediately when a transaction enters the database in order to have much faster results.

3.1.1 Pre-statistical Analysis

When a new transaction enters the system two very important statistical parameters have to be identified: (a) how many items the transaction has and (b) classify each item of the transaction. The first step is trivial since we just have to count how many items the transaction has. When a new transaction enters the database we increase the index of the relative transaction size in a separate table. The second has to do with the fact the LERP-RSA data structure allows classification and parallelism and we want to use them both in a later stage to accelerate the analysis phase. For the classification we just have to detect the initial character or characters of each item in the transaction. Then we increase by one in a separate table the index of each class for the specific item identified. For example, we may have one hundred items and we use the decimal digits to denote each one (i.e., item symbolism starting at "00" and ending at "99"). When the item, e.g., 23 is found in the transaction then the index of class 2 is increased by one, if we use Level 1 Classification (as described in [6]), or the index of the class 23 is increased by 1, if we use Level 2 Classification. The classification process is very important as we can observe in [6] because it allows the partition of the database to very small tables that can be analyzed in parallel, as we will see in step 4.

When all the transactions have been inserted in the database we should have two small tables showing, first how many transactions exist per items number (i.e., how many single item transactions, two items transactions, etc.) and second how many occurrences of items per class exist (i.e., how many items starting with the first digit, the second, etc.). The sum of the first table must have the same value as the total number of transactions recorded, since we have just grouped the transaction based on how many items they have. The sum of the second table shows how many items occurred in total in all transactions imported to the database. For example, if we have 150 transactions in total partitioned in 100 transactions of one item and 50 of two items, the first table should hold exactly this information and the sum should be 150 while the second table should hold the occurrences per alphabet for the item and the sum should be 200. The total computation cost for this phase is non-measurable and we cannot take it in consideration as a friction for the process.

3.1.2 Transaction String Transformation

Since we want to use the LERP-RSA data structure and apply the ARPaD algorithm we have to transform each transaction to a string. However, this is not something that can be done directly as the transaction enters the system. Suppose, for example, that we have 9999 items and the transaction {1, 11, 111, 1111} enters the system. If we just concatenate the transaction like 1111111111, then we have completely lost the information because the specific string can mean any kind of transaction (except the original) like {11, 11, 111, 1} or {1, 1111, 111}, etc. In order to avoid this problem we have to determine how many items could exist for the specific transactions' category database. This is important because if we want to number items by assigning integers to each item, then we have to know up to which integer we can go in order to know how many digits the top most item can have. For example, we may have 1000 items from 0 up to 999 or 50,000 items from 0 up to 49,999. In the first case we care about strings per item with length three while in the second with length five.

After determining how many possible items we can have we apply the following technique. If we use integers in their decimal system representation, then our alphabet is {0, 1, 2, 3, 4, 5, 6, 7, 8, 9}. In this alphabet we will add a neutral symbol, e.g., $ and the alphabet will become {0, 1, 2, 3, 4, 5, 6, 7, 8, 9, $}. Then for every item in the transaction with number of digits less than the possible top most item can have, we add at the end as many $ are needed in order to make all items in the transaction to have the same length as strings. In the previous example our string for the transaction {1, 11, 111, 1111} instead of 11111111 it will become 1$$$11$$111$1111. All transactions' strings after this transformation will have length of a multiple of the number of digits the possible top-most item can have (in this case a multiple of 4). This process can be generalized for any kind of number of items and any alphabet (e.g., hexadecimal). In general this transformation does not aggravate significantly the size of the string since even if we have up to one billion items, then we need in worst case a multiple of 9. Again the total computation cost for this process is non-measurable and cannot be taken into consideration.

The use of the neutral symbol is extremely important for two reasons: space and speed. Although it seems that the use of extra symbol, which creates longer substrings for items with small numbering, can be an overhead and needs more space and time for analysis, this is not true. Let's explain this by assuming that we have in total 1,000,000 items and a transaction like, e.g., {1, 11, 111, 1111, 11111, 111111}. If we use a delimiter to create the corresponding string we will get 1,11,111,1111,11111,111111 while with the use of the neutral symbol the transaction will be transformed to the string 1$$$$$11$$$$111$$$1111$$11111$111111. The first string with the delimiter has length 19 while the second with the neutral symbol has length 25 and, therefore, is longer. However, if we assume that every item has the same probability to occur in a transaction, then the single digit numbered items have probability 10/1,000,000 or 1 out of 100,000, items with double digit have probability 90/1,000,000 or 9 out of 100,000, three digits items have probability 900/1,000,000 or 9 out of 10,000, four digits items have probability 9,000/1,000,000 or 9 out of 1000, five digits items have probability 90,000/1,000,000 or 9% while six digits items have probability 900,000/1,000,000 or 90%. This is very important when considering space for the strings and we can show it with the following example. Let's assume that we have a transaction of all items. In this case using the delimiter we need 10 bytes for single digit, 180 for double digit, 2700 for three digits, 36,000 for four digits, 450,000 for five digits, 5,400,000 for six digits, and we also need 999,999 delimiters. The sum of all these is 6,888,889 bytes in total. However, if we use the neutral symbol we need 6 bytes for each item out of the 1,000,000 or exactly 6,000,000 bytes which needs almost 15% less space. In general the use of the neutral symbol is much more convenient and less space consuming because the 90% of the items have length 6 and there is no need to use the neutral symbol while with the delimiter the use of it cannot be avoided. Therefore, in the majority of transactions the space we need with the neutral symbol is less than the use of delimiter. Only when we have transactions with low numbered items the delimiter can preserve space comparing to neutral symbol, yet, even in the case we can assign low digits numbers to items that do not occur frequently and avoid this problem. In general we can observe that for the 90% of the cases we do not need to use the neutral symbol while the delimiter has to be used for these cases, while for the 9% of the cases we do have to use an extra byte for the both methods: the neutral symbol or the delimiter. Only for the 1% of the cases the use of the delimiter can perform better than the neutral symbol, which allows the neutral symbol to outperform comparing to delimiters.

The next important factor of the neutral symbol is speed. Assuming the previously described example of the 1,000,000 items transaction and given the fact that the items will rather occur in a random order than in ascending order, therefore, in order to read the transaction we have to read the whole string digit by digit because simply we do not know when a delimiter occurs. In this case we need 6,888,889 steps in order to decompose the string to the original items and check for frequent itemsets. With the use of the neutral symbol though this can be avoided. Since every item has length 6 digits, with the use of the neutral symbol for items numbered with less than 6 digits, we have just to jump every 6 digits in order to decompose the string to the original items and detect the frequent itemsets and, therefore, be 6–7 times approximately faster. Furthermore, the use of the neutral symbol gives us a significant advantage when we want to use wildcards, i.e., having gaps between the items' order or in case we do not care about which item occurs in a specific position in an itemset. For example, we may have itemsets of three items starting with item 111 and ending with item 333, without caring which one is in between. In such case having a string of type 111$$$xxxxxx333$$$ is more convenient for analysis purposes and it is definitely needed by the algorithm since after reading the first item at position 1 ARPaD can jump directly to position 13 where the third item occurs without knowing what is in between. ARPaD will always know that the hopping step is six in the above-mentioned example. If we use a delimiter, then having an itemset 111,2,333 is completely different than the itemset 111,222,333 and we have to read the whole string digit by digit. This property of the methodology can also be used during the meta-analyses processes to produce more complicated queries with the use of wildcards and, therefore, more advanced and significant results.

3.1.3 LERP-RSA Construction

After completing recording all transactions in the database, the next important step for our process is the construction of the LERP-RSA suffix array. In this case we have to take the string constructed in the previous phase and create and store all suffix strings. However, we can significantly reduce the number of suffix strings we have to create by taking into consideration that all important suffix strings should have length multiple of the size of the larger item. In the previous example with the string 1$$11$2$$222, although it has length 12 and we should create 12 suffix strings (including the original string) we can just create as many as the items in the transaction, i.e., four (1$$11$2$$222, 11$2$$222, 2$$222, and 222). Every other suffix string does not correspond to an item or group of items and, therefore, it is not important. This will not just limit the required space we need to store all suffix strings but it will also accelerate the analysis process since less strings will have to be analyzed. It has to be mentioned that if instead of the neutral symbol we use a delimiter, then the above process cannot be completed and we have to create all suffix strings causing significant problems during the data mining step.

However, as we have mentioned before, in most cases we do not care for this process to apply ARPaD algorithm in order to detect all repeated patterns but just the patterns that have support (frequency) above a specific threshold. For this we can apply two techniques. The first one has to do with the statistics we have calculated in step 1. Since we care about sequential transaction this means that we care about the arrangement of the items, i.e., {1, 2} is different than {2, 1}. In this case the number of items the frequent itemsets can have depends on how the items appeared (arranged) in each transaction. Therefore, the number of items we can have in a frequent itemsets and has occurred above a specific threshold is directly dependable on the number of transactions with items equal or more than this number. For example, if more than half of the transactions have one or two items, then it is impossible to have itemsets of three or more items with support 50%. Therefore, in the example with the string 1$$11$2$$222 instead of creating the suffix strings 1$$11$2$$222, 11$2$$222, 2$$222, and 222 we can create and store sub-suffix strings of up to, e.g., two items, i.e., 1$$11$, 11$2$$, 2$$222, and 222 which correspond to the original suffix string but truncated to have at most two items. This technique further improves the required space needed for the LERP-RSA data structure and it will also improve the required time as we will observe with the experimental analysis. Of course the last suffix string (211) does not need to be stored since we care about sequential itemsets and just one item does not comply with this, unless we want to detect also the frequency of each discrete item, something that ARPaD algorithm can do for us very easily.

The second technique derives from the Theorem 2 of [10] (p. 9), which briefly states that for a given string the probability to have repeated patterns in this string that are very long is extremely small:

Theorem (Probabilistic Existence of Longest Expected Repeated Pattern) Let S be a string of size n constructed from a finite alphabet Σ of size $m \geq 2$ and let s be a substring (of S) of length l. Let X be the event that "substring s occurs at least twice in string S." If S is considerably long and random, and s is reasonably long, then the probability $P(X)$ of the event X is extremely small. An upper bound for the probability $P(X)$ is directly proportional to the square of the length of string S and inversely proportional to the size of the alphabet m raised to the power of the length of substring s:

$$\overline{P(X)} = \frac{(n - 2 * l + 1) * (n - 2 * l + 2)}{2 * m^l} < \frac{n^2}{2 * m^l}$$

or

$$\overline{P(X)} = \frac{n^2}{2 * m^l}$$

where the over-bar is used to define an upper bound of the underlying value.

Therefore, if we consider the concatenation of all strings of transactions as one large string, then the probability to have very long patterns (itemsets) is extremely small. This means that although for a given support we can have, for example, up to ten items per frequent itemset based on the statistics of step 1, however, based on the Theorem, the probability to have such long strings (ten items per frequent itemset) is extremely small and, therefore, it will end to be less than the desired support. According to the theorem we are safe to assume and expect frequent itemsets, which will have support above the specific threshold, with less than ten items. Lemma 2 in [6] (p. 16) allows estimating the length of the patterns (in our case the number of items per frequent itemset) with a very high probability:

Lemma 2 Let S be a random string of size n, constructed from a finite alphabet Σ of size $m \geq 2$, and an upper bound of the probability $P(X)$ is $\overline{P(X)}$, where X the event "LERP is the longest pattern that occurs at least twice in S." An upper bound for the length l of the Longest Expected Repeated Pattern (LERP) we can have with probability $P(X)$ is:

$$\bar{l} = \frac{\log \frac{((n-2*l+1)*(n-2*l+2))}{2*P(X)}}{\log m} < \left\lceil \frac{\log \frac{n^2}{2*P(X)}}{\log m} \right\rceil$$

or

$$\bar{l} = \left\lceil \frac{\log \frac{n^2}{2*P(X)}}{\log m} \right\rceil$$

where $l \ll n$ and $\overline{P(X)} > 0$.

Therefore, we can further reduce the size of the sub-suffix strings that we have to store in the database and simultaneously reduce the overall space we require to store the suffix strings and the time for the analysis. This also helps with the sorting process since after crating all suffix strings we have to lexicographically sort them. If suffix strings are longer it is obvious that any sorting algorithm needs additional processing time in order to complete the process.

Before start storing the suffix strings in the database we have also to determine the Classification Level we will apply to our data structure [6]. LERP-RSA, as we mentioned earlier, allows the classification of the suffix strings according either (a) to the first character of the string or (b) to the arrangements of characters of the alphabet. So, we can apply Level 1 Classification over the decimal alphabet and creating ten classes, one for each digit, Level 2 Classification and creating one hundred classes (from "00" up to "99"), etc. The factor that will guide us to determine what kind of Classification Level we will apply is the second statistical observation we have committed in the first phase by classifying all items occurred in every transaction. If the sizes of the classes are equidistributed, then we can apply Classification Level 1, however, if one or more classes have significantly larger size, then for these classes we can apply Classification Level 2. This process can be continued to other Levels of Classification depending on how the sizes of the classes are distributed. This step is very important for the next phase.

3.2 ARPaD Data Mining Phase

The second phase of the method starts when all transactions have entered the database and the corresponding LERP-RSA tables have been constructed. ARPaD algorithm will be used to detect all repeated patterns and the results of the data mining can be further examined in meta-analyses in order to extract useful information. ARPaD algorithm also allows the use of shorter pattern length (SPL), which is not important in this case unless we care for itemsets having more items than a specific threshold. ARPaD algorithm works as follows [5] (p. 951):

1. For all the letters of the alphabet, count suffix strings that start with the specific letter.
2. If no suffix strings are found or only one is found, proceed to the next letter (periodicity cannot be defined with just one occurrence).
3. In case the same number of substrings is found as the total number of the suffix strings, proceed to step 4 and the specific letter is not considered as occurrence because a longer hyper-string will occur.
4. If more than one string and less than the total number of the suffix strings is found, then for the letters used and counted already and for all letters of the alphabet add a letter at the end and construct a new hyper-string. Then do the following checks:

   a. If none or one suffix string is found that starts with the new hyper-string and the length of the previous substring is equal to or larger than SPL and smaller than LERP, consider the previous substring as an occurrence, find previous substrings' positions and proceed with the next letter of the alphabet.

b. If the same number of substrings is found as previously and the length of the previous substring is smaller than LERP, then proceed to step 4. However, the specific substring is not considered as occurrence because a longer hyper-string will occur.

c. If more than one and less than the number of occurrences of the previous substring is found and the length of the previous substring is different than LERP, consider the previous substring as a new occurrence. If the previous substring has not been calculated again and the length of the substring is equal to or longer than SPL, then calculate substrings' positions. Continue the process from step 4.

Algorithm. All Repeated Patterns Detection with Shorter Pattern Length and Longest Expected Repeated Pattern

Input: String of pattern we want to check, a counter of string length, SPL length, LERP length

Output: A list of all occurrence vectors

1	ARPaD_LERP (string X, int count, int SPL,int LERP)
2	isXcalculated := false
3.1	for each letter l in alphabet
3.2	newX := X + l
3.3	newCount := how many strings start with newX
3.4.1	if (newCount = count) AND (X.length < LERP)
3.4.2	ARPaD_LERP (newX, newCount, SPL,LERP)
3.4.3	end if
3.5.1	if (count > 1) AND (X.length = LERP) AND (isXcalculated = false)
3.5.2	find positions of string X
3.5.3	isXcalculated := true
3.5.4	end if
3.6.1	if (newCount = 1) AND (isXcalculated = false) AND (X NOT null) AND (X.length >= SPL)
3.6.2	find positions of string X
3.6.3	isXcalculated := true
3.6.4	end if
3.7.1	if (newCount > 1) AND (newCount < count) AND X.length <> LERP)
3.7.2.1	if (isXcalculated = false) AND (X NOT null)
3.7.2.1	AND (X.length >= SPL)
3.7.2.2	find positions of string X
3.7.2.3	isXcalculated := true
3.7.2.4	end if
3.7.3	ARPaD_LERP (newX, newCount, SPL, LERP)
3.7.4	end if
3.8	end for
4	end ARPaD_LERP

3.2.1 Frequent Sequential Itemsets Detection

After completing the LERP-RSA data structure creation phase, we run the ARPaD algorithm on the database in order to detect all repeated patterns, which in our case are frequent and non-frequent itemsets. We have mentioned that LERP-RSA data structure allows the parallel execution of the ARPaD algorithm. Since we have applied classification in the previous phase and we have many tables we can run ARPaD in parallel for each class. Yet, here the step depends on the hardware we have and the number of classes. If, for example, the CPU allows up to ten threads in parallel and we have ten classes, we can apply each class to one thread and run ARPaD. This is the "Static Parallel" execution of the algorithm for frequent itemsets detection. However, we can take advantage of the statistics done in step 1 and the number of classes created in step 3, in order to apply "Dynamic Parallel" execution, as follows. If the classes are equidistributed we can simply execute one thread per class for the algorithm and the total time needed for the analysis it will be the time of the largest class and, therefore, slowest to be analyzed. Nevertheless, if the classes are not equidistributed, then we can start with one thread per class and when ARPaD finishes for one class we can use the dismissed thread to execute the algorithm to another class. In this case there will be no idle thread and processing time for the CPU and the whole process can finish significantly faster as we will observe for the experimental results in the next section.

3.2.2 Meta-Analyses of the Results

After the execution of ARPaD algorithm and the detection of all repeated patterns (i.e., itemsets) we have to do a meta-analysis of the results. The results returned from the ARPaD are extreme in number since the algorithm does not take into consideration support (does not take such input argument) and it will return everything that occurred at least twice. This means that we will have itemsets with extremely small support. This though is not a problem since we can run any kind of query on the results stored in the database in order to find exactly the itemsets we care most. This actually is a very good attribute of the algorithm because it allows us to run the analysis once and then run as many meta-analyses as we want directly on the results. For example, we can run query such as how many items have support in between specific thresholds and/or include specific items, etc. In general we have two main categories of meta-analyses that can be conducted:

- Item specific: multiple queries can be run in the results to detect specific items or groups of items. This kind of analyses can be used to check how different items or groups are interchangeable and/or are correlated.
- Occurrences specific: multiple queries can be run in order to detect trends in items or groups of items. This kind of analyses are very useful in order to detect how itemsets occurrences fluctuate over time. Therefore, it can be observed that low frequency itemsets have a rapid growth or tend to decrease and eventually will be important itemsets, for e.g., marketing purposes.

More advanced queries can be executed combining characteristics from both categories and giving more complicated results, important for business intelligent. The results can also be stored in a data warehouse and compared with newer results from analyses conducted much later, etc. The possibilities of such analyses are endless.

In order to make clear how SAFID method and ARPaD algorithm work, the following example will be presented.

Example 1 For the example we will use seven transactions from the "Retail" dataset which can be found on FIMI website. The seven transactions are (1) no. 27 {39, 41, 48}, (2) no. 62 {32, 39, 41, 48, 348, 349, 350}, (3) no. 85 {32, 39, 41, 48, 152, 237, 396}, (4) no. 172 {39, 41, 48, 854}, (5) no. 193 {10, 39, 41, 48, 959, 960}, (6) no. 528 {38, 39, 41, 48, 286}, and (7) no. 550 {39, 41, 48, 89, 310}. First of all we have to observe that the top-most item has number 960 (three digits length) and, therefore, we need to make all items of the same size, i.e., three. So, when the first transaction enters the system first we have to detect how many items have (three) and classify these items (one for class 3 and two for class 4). Then the transaction has to be transformed to string 39$41$48$. We do the same for all transactions and we have the following results for length (Table 1) and classification (Table 2).

The first table sums up to seven which is the number of transactions while the second to 37 which is the total number of items in all transactions. Moreover, the new transactions have been transformed to strings 32$39$41$48$348349350, 32$39$

Table 1 Lengths

Length	Transactions
1	0
2	0
3	1
4	1
5	2
6	1
7	2

Table 2 Items classification

Class	Items
0	0
1	2
2	2
3	15
4	14
5	0
6	0
7	0
8	2
9	2

(a)	(b)	(c)
39$41$48$	10$39$41$48$959	10$39$41$48$959
41$48$	152237396	152237396
48$	237396	237396
32$39$41$48$348349350	286	286
39$41$48$348349350	310	310
41$48$348349350	32$39$41$48$152	32$39$41$48$152
48$348349350	32$39$41$48$348	32$39$41$48$348
348349350	348349350	348349350
349350	349350	349350
350	350	350
32$39$41$48$152237396	38$39$41$48$286	38$39$41$48$286
39$41$48$152237396	396	396
41$48$152237396	39$41$48$	39$41$48$
48$152237396	39$41$48$152237	39$41$48$152237
152237396	39$41$48$286	39$41$48$286
237396	39$41$48$348349	39$41$48$348349
396	39$41$48$854	39$41$48$854
39$41$48$854	39$41$48$89$310	39$41$48$89$310
41$48$854	39$41$48$959960	39$41$48$959960
48$854	41$48$	41$48$
854	41$48$152237396	41$48$152237396
10$39$41$48$959960	41$48$286	41$48$286
39$41$48$959960	41$48$348349350	41$48$348349350
41$48$959960	41$48$854	41$48$854
48$959960	41$48$89$310	41$48$89$310
959960	41$48$959960	41$48$959960
960	48$	48$
38$39$41$48$286	48$152237396	48$152237396
39$41$48$286	48$286	48$286
41$48$286	48$348349350	48$348349350
48$286	48$854	48$854
286	48$89$310	48$89$310
39$41$48$89$310	48$959960	48$959960
41$48$89$310	854	854
48$89$310	89$310	89$310
89$310	959960	959960
310	960	960

Fig. 2 LERP-RSA (**a**), lexicographically sorted LERP-RSA (**b**) and ARPaD results (**c**)

41$48$152237396, 39$41$48$854, 10$39$41$48$959960, 38$39$41$48$286, and 39$41$48$89$310, respectively. As we can observe in the first case more space is consumed with the neutral symbol, in fourth, sixth, and seventh case the space is exactly the same while in the other two cases less space has been consumed comparing to the use of a delimiter. The final step is to create the LERP-RSA data structure for the strings "Fig. 2a," lexicographically sort it "Fig. 2b" and execute ARPaD algorithm on it to detect all sequential frequent itemsets.

From the pre-statistical analysis we can observe that we have one transaction of length three, one of length four, two of length five, one of length six, and two of length seven. Therefore, if we care to detect sequential frequent itemsets of frequency greater than 50% there is no need to search for itemsets of size six or seven because they cannot exist (only 3 out of 7). Therefore, suffices of the second, third, and fifth strings do not need to be calculated for length more than five and can be truncated (underlined parts in red color "Fig. 2a"). After this process the lexicographically sorted LERP-RSA has the form of "Fig. 2b".

The final step now is to execute ARPaD algorithm and detect all sequential frequent itemsets. ARPaD starts with the first digit "1" and finds two suffix strings and continues by building longer strings. However, the "10" exists only once and, therefore, the process stops for "1" because it cannot exist more than once for the second substring starting with "15". The second digit is "2" and we have the same situation like "1" with only two substrings. Again the process stops and ARPaD continues with the next digit "3" and finds 15 different substrings. Since there are more than one, ARPaD continues by constructing longer substrings and checks for occurrences of suffix strings starting with 30, 31, 32, etc., up to 39. During this process a substring found with more than one occurrences, i.e., "32$" occurring 2 times. Then the algorithm continues by constructing recursively longer substrings (e.g., 32$39$, etc.) and finds that 32$39$ occurs again 2 times. It continues recursively to detect longer substrings which might occur more than once and stops when the "32$39$41$48$" has been found. The next significant occurrence of substrings (i.e., more than once) is 39$ and for this ARPaD detects the substring "39$41$48$". The algorithm then terminates for this digit and continues with "4" for which we have seven occurrences of substring 41$ and 48$. For the first detects the substring "41$48$" while for the second we have a single item substring 48$. ARPaD algorithm continues and detects only two more occurrences for substrings starting with digits "8" and "9" without having more than one occurrences each digit and it terminates.

Finishing, ARPaD algorithm has detected four itemsets, namely {32, 39, 41, 48}, {39, 41, 48}, {41, 48}, and {48} "Fig. 2c," occurring 2, 7, 7, and 7 times, respectively. These results encapsulate every shorter itemset which exist in the original detected itemsets, e.g., {39, 41} etc. Instead of repeating the process digit by digit we can assume that each item is part of the alphabet and execute ARPaD with a greater alphabet size. Although this seems more efficient, yet, it can slow down the process when the number of items has size hundreds or more.

So far we have not used one parameter of the pre-statistical analysis and this is classification. We can observe that we have approximately half of the suffix strings starting with "3" and "4" (15 and 14 occurrences, respectively) and only two for "1", "2," and "8". Therefore, we can apply parallelism by assigning more threads for the classes "3" and "4" one for the other three classes ("Static Parallel" execution) or assign one thread per digit and when classes "1", "2," and "4" finish assign the dismissed threads to classes "3" and "4" ("Dynamic Parallel" execution). This approach can significantly speed up the process as we have discussed in Sect. 3.2.1 and we will observe from the experimental analysis of the complete datasets in the following section.

4 Experimental Analysis

For the experimental analysis of our methodology we have used a standard desktop computer with an Intel i5 processor at 2.8 GHz with four cores, 8 GB RAM and a 200 GB hard disk. An external DBMS Microsoft SQL Server 2012 has also been used to support the analysis, i.e., database storage. We have executed in total ten experiments on two datasets from the FIMI website (http://fimi.ua. ac.be/data/). The first dataset we have used is the "Retail" and the second is the "Kosarak." The difference of the two datasets are (a) the bigger size of Kosarak comparing to Retail since it has more transactions, items, and items per transaction on average and (b) Kosarak's transactions seem to have some kind of structure. We have conducted six experiments for the first dataset for different support values in static and dynamic parallel execution of the ARPaD algorithm. For the second dataset we have conducted four experiments for two different support values and again in static and dynamic parallel execution. ARPaD algorithm in all experiments used full analysis by detecting even single items. Using the full implementation of the algorithm, yet, the proposed methodology is extremely fast. However, a direct comparison to literature cannot be made because of the fact that our methodology discovers all repeated itemsets regardless their frequency. A direct comparison requires every other method in literature to have detected every frequent itemset regardless its support value. To the best of our knowledge, such analysis does not exist in literature so far. Furthermore, our attempts of testing other methods using well-known academic open source software failed due to crashes, insufficient hardware, or long waiting periods when used low support thresholds that can approach our method's capabilities and results.

4.1   Experiments with "Retail" Dataset

As we can see from Table 3 the first dataset has 88,162 transactions, 16,469 items in total, and we had to create 820,414 records for the LERP-RSA. The sizes of the transactions vary from 1 up to 76 (Table 4). For all three experiments with support 0% (all duplicate itemsets detected) 20 and 80% the sorting time for the LERP-RSA was less than a second (Table 4). Moreover, we can observe from Table 4 the significant improvement of the overall time from the Static Parallel execution to the Dynamic Parallel execution. We can also observe the huge amount of itemsets ARPaD algorithm detected and are more than 120,000 in all cases. Although the analysis may seem to be slow as we can see from the average seek time per itemset it is actually extremely fast. The overall time is affected by the number of itemsets found and this can change by reducing the maximum items per itemset factor while creating the LERP-RSA. As we can see from Table 6 itemsets of size four items or less can have support more than 80% while itemsets of size 16 or less can have support 20%. Yet, in order this support to be present, itemsets in all transactions of this size should exist, something which is extremely difficult. From Table 5 we can see how the sizes of the classes are distributed. We can observe that there is an inequality between classes, with class 1 having almost one third of the itemsets and classes 2, 3, and 4 approximately 15%, something that we can take advantage and execute ARPaD in Dynamic Parallel. In Table 7 we can see an indicative list of the first 20 itemsets with the higher support, regardless of any support factor. As we can observe ARPaD algorithm has also identified single items. In Table 8 we

Table 3 Database size

Transactions	Items	LERP-RSA records
88,162	16,469	820,414

Table 4 Execution time

Frequency (%)	Maximum items per itemset	Sorting (s)	ARPaD (s) Static parallel	Dynamic parallel	Total itemsets found Static parallel	Average seek time (s) Dynamic parallel	
0	76	<1	1471	937	127,196	0.0116	0.0074
20	16	<1	1434	834	127,196	0.0113	0.0066
80	4	<1	1378	730	122,280	0.0113	0.0060

Table 5 Classification per alphabet digit

Class	Occurrences	Frequency (%)
0	177	0.02
1	254,484	28.01
2	131,192	14.44
3	168,656	18.56
4	136,898	15.07
5	60,452	6.65
6	47,415	5.22
7	40,478	4.46
8	36,770	4.05
9	32,054	3.53

can see the total itemsets found per itemset size and we can observe that the longest repeated itemset that exists twice in the transactions' database has size 12 and is the sequential itemset {474, 1246, 1337, 2426, 2741, 3028, 3699, 5125, 5789, 7041, 7111, 7893} which exists at transactions numbered 7164 and 7168 in the dataset (Tables 3, 4, 5, 6, 7, and 8).

4.2 Experiments with "Kosarak" Dataset

As we can see in Table 9, the second dataset has 990,002 transactions, approximately 41,270 items in total and we had to create 7,029,013 records for the LERP-RSA. The sizes of the transactions vary from 1 up to 2498 (Table 12). For both experiments with support 20 and 80% the sorting time for the LERP-RSA was approximately 5 s (Table 10). Moreover, we can observe from Table 10 the significant improvement of the overall time from the Static Parallel execution to the Dynamic Parallel execution. We can also see the vast amount of itemsets ARPaD

Table 6 Top 20 and bottom 10 sequential frequent sizes

Size	Occurrences	Frequency (%)	Reverse cumulative frequency (%)
1	3,016	3.42	100
2	5,516	6.26	96.58
3	6,919	7.85	90.32
4	7,210	8.18	82.47
5	6,814	7.73	74.3
6	6,163	6.99	66.57
7	5,746	6.52	59.58
8	5,143	5.83	53.06
9	4,660	5.29	47.23
10	4,086	4.63	41.94
11	3,751	4.25	37.31
12	3,285	3.73	33.05
13	2,866	3.25	29.32
14	2,620	2.97	26.07
15	2,310	2.62	23.1
16	2,115	2.4	20.48
17	1,874	2.13	18.08
18	1,645	1.87	15.96
19	1,469	1.67	14.09
...	...	...	...
63	5	0.01	0.03
64	2	0	0.02
65	2	0	0.02
66	5	0	0.02
67	3	0	0.01
68	3	0	0.01
71	1	0	0
73	1	0	0
74	1	0	0
76	1	0	0

algorithm detected and are more than 1.6 million in the first case and almost 370 thousand in the second. Although the analysis may again seem to be slow as we can observe from the average seek time per itemset it is again extremely fast. The overall time is affected by the number of itemsets found and this can change by reducing the maximum items per itemset factor while creating the LERP-RSA. As we can see from Table 12 itemsets of size two items or less can have support more than 80% while itemsets of size seven or less can have support 20%. Again as with the previous dataset, in order this support to be present itemsets in all transactions of this size should exist. From Table 11 we can see how the sizes of the classes are distributed. We can observe that again we have an inequality between classes, with class 1 having one quarter of the itemsets and 2, 3, and 6 having approximately 15%,

Table 7 Top 20 more frequent sequential itemsets

Itemset	Items per itemset	Occurrences	Frequency (%)
39	1	49,618	56
48	1	41,178	47
39, 48	2	21,153	24
38	1	15,534	18
32	1	14,888	17
41	1	14,756	17
39, 41	2	11,386	13
38, 39	2	10,345	12
41, 48	2	8,775	10
39, 41, 48	3	7,146	8
32, 39	2	6,591	7
65	1	4,427	5
38, 39, 48	3	3,853	4
89	1	3,765	4
225	1	3,192	4
38, 39, 41	3	3,039	3
237	1	2,989	3
36	1	2,934	3
32, 39, 48	3	2,903	3
170	1	2,894	3

Table 8 Itemsets per itemset size

Itemset size	Itemsets
1	12,217
2	80,399
3	20,890
4	8,774
5	3,539
6	1,079
7	227
8	48
9	14
10	6
11	2
12	1

something which again we used in order to execute ARPaD in Dynamic Parallel. In Table 13 we can see an indicative list of the first 20 itemsets with the higher support, regardless of any support factor. As we can observe ARPaD algorithm has also identified single items with high frequency. In Table 14 we can see the top 20 categories of distinct itemsets occurred per itemset size. We can witness from the list of Table 14 the vast amount of itemsets exist for different itemsets' sizes and, despite

Table 9 Database size

Transactions	Items	LERP-RSA records
990,002	41,270	7,029,013

Table 10 Execution time

Frequency (%)	Maximum items per itemset	Sorting (s)	ARPaD (s)		Total itemsets found	Average seek time (s)	
			Static parallel	Dynamic parallel	Static parallel	Dynamic parallel	
20	7	5	20,728	12,363	1,648,282	0.0126	0.0075
80	2	5	3,389	2,067	372,109	0.0091	0.0056

Table 11 Classification per alphabet digit

Class	Occurrences	Frequency (%)
0	0	0
1	2,120,866	26.45
2	1,227,279	15.3
3	1,268,076	15.81
4	725,381	9.05
5	573,701	7.15
6	1,034,064	12.9
7	481,290	6
8	342,751	4.27
9	245,607	3.06

that, ARPaD has managed to analyze and detect all repeated itemsets regardless of their frequency. Although having itemsets that exist few times may seem of no value, yet, it allows the in-depth analysis of the transactions and the detection of more complex patterns and combinations of them that can exist in the results by re-analyzing directly the results, which has low computational cost, instead of the original, initial database of transactions.

It is important to mention that for both datasets, as we can observe from Tables 7 and 13, there are no itemsets with frequency above 80% while there are only three for both cases with frequency above 20%. When we use frequency 20 and 80% in Tables 4 and 9 we mean the theoretical optimum frequency we can have based on the Reverse Cumulative Frequency column of Tables 6 and 11, respectively, as we have discussed in Sect. 3. This optimum frequency is used for the construction of the LERP-RSA in order to minimize the required space while avoiding losing information. However, as mentioned earlier, Theorem 2 [25] guarantees with high probability that all itemsets will exist with frequency much less than the optimum we can have from the Reverse Cumulative Frequency and, therefore, we can further reduce the size of the LERP-RSA and accelerate the analysis by using Lemma 2 [25] to estimate the optimum probabilistic value for the LERP-RSA size (Tables 9, 10, 11, 12, 13, and 14).

Table 12 Top 20 and bottom 10 sequential frequent sizes

Size	Occurrences	Frequency (%)	Reverse cumulative frequency (%)
1	152,796	15.43	100
2	198,372	20.04	84.57
3	173,804	17.56	64.53
4	119,916	12.11	46.97
5	73,735	7.45	34.86
6	45,138	4.56	27.41
7	30,112	3.04	22.85
8	22,001	2.22	19.81
9	17,489	1.77	17.59
10	14,512	1.47	15.82
11	11,922	1.2	14.36
12	10,380	1.05	13.15
13	8,831	0.89	12.1
14	7,835	0.79	11.21
15	6,815	0.69	10.42
16	6,078	0.61	9.73
17	5,414	0.55	9.12
18	4,947	0.5	8.57
19	4,443	0.45	8.07
...	...	...	...
1572	1	0	0
1643	1	0	0
1654	1	0	0
1733	1	0	0
1828	1	0	0
1926	1	0	0
2108	1	0	0
2337	1	0	0
2491	1	0	0
2498	1	0	0

5 Conclusion

In the current paper we proposed a new methodology for the detection of frequent and non-frequent sequential itemsets (SAFID). For this purpose, we have used a new variation of the already developed methodology for the detection of all repeated patterns in a string. The flexibility of the LERP-RSA data structure and the efficiency of the ARPaD algorithm allow the detection of all sequential itemsets in a database of transactions in a very fast time, without taking into consideration any support threshold. Moreover, our methodology has significant advantage since it allows the meta-analyses of the results by applying advanced and complex queries directly on the results and not on the original database. This also allows

Table 13 Top 20 more
frequent sequential itemsets

Itemset	Items per itemset	Occurrences	Frequency (%)
6	1	525,880	53
11	1	358,412	36
3	1	234,770	24
1	1	182,513	18
11, 6	2	120,247	12
6, 3	2	104,716	11
1, 6	2	93,859	9
218	1	86,673	9
4	1	73,174	7
27	1	70,400	7
7	1	69,963	7
218, 6	2	65,233	7
55	1	52,845	5
11, 1	2	50,311	5
148	1	41,007	4
2	1	37,473	4
64	1	35,554	4
11, 1, 6	3	32,304	3
11, 27	2	32,196	3
27, 6	2	32,066	3

the comparative study of different datasets over time and the detection of useful information such as trends on how itemsets' occurrences evolve during time.

Due to the nature of the methodology it is not directly comparable to other methods because it detects not itemsets above a specific frequency (support) but every repeated itemsets, frequent and non-frequent. Yet, regardless the enormous amount of results SAFID returns, it has been proven to be very fast.

Table 14 Top 20 itemsets
per itemset size

Itemset size	Itemsets
1	28,429
2	343,680
3	509,712
4	371,528
5	210,158
6	114,798
7	69,977
8	50,917
9	42,635
10	38,528
11	36,264
12	34,898
13	34,185
14	33,695
15	33,158
16	32,933
17	32,930
18	33,015
19	32,988
20	32,999

References

1. Agrawal R, Srikant R. Mining sequential patterns. In: Yu PS, Chen ASP, editors. 11th International Conference on Data Engineering (ICDE'95). Taipei, Taiwan: IEEE Computer Society Press; 1995. p. 3–14.
2. Srikant R, Agrawal R. Mining sequential patterns: generalizations and performance improvements. Berlin Heidelberg: Springer; 1996. p. 1–17.
3. Akerkar R, Akerkar R. Discrete mathematics. New Delhi: Pearson India; 2007.
4. Enberton H. Set theory. Encyclopedia Britannica. http://www.britannica.com/topic/set-theory.
5. Xylogiannopoulos K, Karampelas P, Alhajj R. Analyzing very large time series using suffix arrays. Appl Intell. 2014;41(3):941–55.
6. Xylogiannopoulos K, Karampelas P, Alhajj R. Repeated patterns detection in big data using classification and parallelism on LERP reduced suffix arrays. Appl Intell. 2016:1–31. doi:10.1007/s10489-016-0766-2.
7. Mabroukeh NR, Ezeife CI. A taxonomy of sequential pattern mining algorithms. ACM Comput Surv. 2010;43(1):41.
8. Gupta M, Han J. Applications of pattern discovery using sequential data mining. In: Pattern discovery using sequence data mining: applications and studies. Hershey: Information Science Reference; 2012. p. 1–23.
9. Mooney CH, Roddick JF. Sequential pattern mining—approaches and algorithms. ACM Comput Surv (CSUR). 2013;45(2):19.
10. Masseglia F, Cathala F, Poncelet P. The psp approach for mining sequential patterns. In: Principles of data mining and knowledge discovery. Berlin Heidelberg: Springer; 1998. p. 176–84.

11. Garofalakis MN, Rastogi R, Shim K. SPIRIT: sequential pattern mining with regular expression constraints. VLDB. 1999;99:7–10.
12. Zhang M, Kao B, Yip CL, Cheung D. A GSP-based efficient algorithm for mining frequent sequences. In: Proceedings of IC-AI, June 2001, p. 497–503.
13. Luo C, Chung SM. A scalable algorithm for mining maximal frequent sequences using a sample. Knowl Inf Syst. 2008;15(2):149–79.
14. Zaki MJ. Efficient enumeration of frequent sequences. In: Proceedings of the seventh international conference on Information and knowledge management. New York: ACM; Nov 1998. p. 68–75.
15. Ayres J, Flannick J, Gehrke J, Yiu T. Sequential pattern mining using a bitmap representation. In: Proceedings of the eighth ACM SIGKDD international conference on Knowledge discovery and data mining. New York: ACM; July 2002. p. 429–35.
16. Yang Z, Wang Y, Kitsuregawa M. LAPIN: effective sequential pattern mining algorithms by last position induction for dense databases. In: Advances in databases: concepts, systems and applications. Berlin Heidelberg: Springer; 2007. p. 1020–3.
17. Orlando S, Perego R, Silvestri C. A new algorithm for gap constrained sequence mining. In: Proceedings of the 2004 ACM symposium on Applied computing. New York: ACM; Mar 2004. p. 540–7.
18. Savary L, Zeitouni K. Indexed bit map (ibm) for mining frequent sequences. In: Knowledge discovery in databases: PKDD 2005. Berlin Heidelberg: Springer; 2005. p. 659–66.
19. Han J, Pei J, Mortazavi-Asl B, Chen Q, Dayal U, Hsu MC. FreeSpan: frequent pattern-projected sequential pattern mining. In: Proceedings of the sixth ACM SIGKDD international conference on knowledge discovery and data mining. ACM; Aug 2000. p. 355–9.
20. Pei J, Han J, Mortazavi-Asl B, et al. Prefixspan: mining sequential patterns efficiently by prefix-projected pattern growth. In: 2013 IEEE 29th International Conference on Data Engineering (ICDE). IEEE Computer Society; 2001. p. 0215.
21. Seno M, Karypis G. Lpminer: an algorithm for finding frequent itemsets using length-decreasing support constraint. In: Proceedings of IEEE International Conference on Data Mining, 2001 (ICDM 2001). IEEE; 2001. p. 505–12.
22. Kum HC, Pei J, Wang W, Duncan D. ApproxMAP: approximate mining of consensus sequential patterns. In: SDM. May 2003. p. 311–5.
23. Song S, Hu H, Jin S. HVSM: a new sequential pattern mining algorithm using bitmap representation. In: Advanced data mining and applications. Berlin Heidelberg: Springer; 2005. p. 455–63.
24. Chiu DY, Wu YH, Chen AL. An efficient algorithm for mining frequent sequences by a new strategy without support counting. In: Proceedings of 20th IEEE international conference on data engineering, 2004. Mar 2004. p. 375–86.
25. Yin J, Zheng Z, Cao L. USpan: an efficient algorithm for mining high utility sequential patterns. In: Proceedings of the 18th ACM SIGKDD international conference on Knowledge discovery and data mining. New York: ACM; Aug 2012. p. 660–8.
26. Zihayat M, Wu CW, An A, Tseng VS. Mining high utility sequential patterns from evolving data streams. In: Proceedings of the ASE big data and social informatics 2015. New York: ACM; Oct 2015. p. 52.
27. Ukkonen E. Maximal and minimal representations of gapped and non-gapped motifs of a string. Theor Comput Sci. 2009;410(43):4341–9.
28. Xylogiannopoulos K, Karampelas P, Alhajj R. Experimental analysis on the normality of π, e, φ, sqrt(2) using advanced data-mining techniques. Exp Math. 2014;23(2):105–28.

Index

© Springer International Publishing AG 2017
M. Kaya et al. (eds.), *From Social Data Mining and Analysis to Prediction
and Community Detection*, Lecture Notes in Social Networks,
DOI 10.1007/978-3-319-51367-6

Printed in the United States
By Bookmasters